Molecular Techniques for Studying Viruses

Zubaida Hassan • Gulfaraz Khan

Molecular Techniques for Studying Viruses

Practical Notes

Springer

Zubaida Hassan
Faculty of Life Sciences, Department of Microbiology
Modibbo Adama University
Yola, Nigeria

Gulfaraz Khan
Department of Medical Microbiology and Immunology
College of Medicine and Health Sciences,
United Arab Emirates University
Al Ain, Abu Dhabi, United Arab Emirates

ISBN 978-981-99-8096-3 ISBN 978-981-99-8097-0 (eBook)
https://doi.org/10.1007/978-981-99-8097-0

This Springer imprint is published by the registered company Springer Nature Singapore Pte Ltd.
The registered company address is: 152 Beach Road, #21-01/04 Gateway East, Singapore 189721, Singapore

Zubaida Hassan: This book is dedicated to my beloved husband, Alhaji Adamu Mohammed Hassan, my children, my parents, and my family for making it all worthwhile.
Gulfaraz Khan: This book is dedicated to my family, my wife, my children, and my late parents for their support, sacrifices, patience, and constant forgiveness for my regular absence from home.

Preface

The last 50 years have witnessed enormous advances in virtually all areas of medicine and life sciences. These advances are particularly evident in molecular biology, genetics, biotechnology, and digital imaging. Some of these advances have revolutionalised laboratory-based methods of studying viruses. These methods include the production of monoclonal antibodies, polymerase chain reaction (PCR), high throughput genomics, transcriptomics, and proteomics. These emerging laboratory advances have had a major impact not only on virology research but also on diagnostic virology. Viruses can now be detected in record time with much greater sensitivity and specificity. Moreover, these innovative approaches are beginning to provide a comprehensive global insight into infectious agents and their surrounding environment. Thousands of genes or cellular components can now be analysed simultaneously from a single sample and topographically mapped to tissue morphology. However, many of these technologies are expensive, highly sophisticated, and require considerable expertise to establish and run. As such, they are out of reach for routine hands-on practical training of university students at the beginning of their scientific research training.

This book is based on the introductory post-graduate practical course '*Molecular Techniques in Viral Pathogenesis*' that we have been teaching at the College of Medicine and Health Sciences, UAE University, for many years. The aim was to provide short practical notes that are simple, easy to follow, and implementable in most university labs. One key aspect in selecting the techniques for inclusion was their feasibility for adoption in routine, low-resource settings. Moreover, we wanted to provide practical introductory notes that could be used for hands-on training of students, both undergraduates and junior postgraduates. Thus, these introductory notes are not meant to be a detailed step-by-step laboratory manual but rather a short hands-on book outlining principles and basic practical notes on commonly used techniques for studying viruses. Throughout the book, we have frequently added helpful tips 'NOTE' to draw the attention of the reader to precautionary advice relevant to the practical step being described. This book should serve as an initial guide on which students could build on as they become more familiar with laboratory methods. Furthermore, in this age of digital technology, having the *Practical Notes* available as an ebook in addition to hard copies would serve our goal most effectively. Finally, we hope that students and teachers will find this book helpful in their practical classes.

Yola, Nigeria
Al Ain, Abu Dhabi, United Arab Emirates

Zubaida Hassan
Gulfaraz Khan

Acknowledgements

We would like to thank all our colleagues and researchers, past and present, in the Viral Pathology Laboratory, College of Medicine and Health Sciences, UAE University. Special thanks are due to Pretty S. Philip (Research Specialist), Dr. Waqar Ahmed (Post-doctoral Fellow at Johns Hopkins School of Medicine, USA), Dr. Asma Hassani (Post-doctoral Fellow at Harvard Medical School, USA), and Dr. Narendran Reguraman (Post-doctoral Fellow at Ohio State University, USA) for their practical advice, help, and critical reading of the contents of this book. We would also like to acknowledge and thank the funding bodies who have been supporting research in our laboratory: UAE University (Zayed Centre-based grant), College of Medicine and Health Sciences (CMHS project grant), and the Al Jalila Foundation (Seed-grant). Zubaida Hassan received the UAE University Fellowship Award for her PhD studies. Zubaida Hassan would also like to thank Modibbo Adama University, Yola, Nigeria, for the sabbatical leave to pursue her PhD. Finally, we would also like to thank the publisher and editors, in particular, Ms. Swati Sharma for her regular support and guidance throughout the publication of this book.

Contents

About the Authors

Zubaida Hassan is a lecturer in the Department of Microbiology, School of Life Sciences, Modibbo Adama University, Yola, Nigeria. She did her bachelor's degree at the Federal University of Technology, Yola, Nigeria, her master's degree at the University Putra Malaysia, and PhD at the United Arab Emirates University. Zubaida's PhD project was on 'Structural impact of Epstein-Barr virus-encoded small RNA-1 on its transport and function'. Zubaida has a few publications in peer-reviewed journals and book chapters. She has also received several international conference scholarship awards from reputable organisations to attend and present her work.

Gulfaraz Khan is a professor of viral pathology in the Department of Medical Microbiology & Immunology, College of Medicine and Health Sciences, UAE University. Dr. Khan received his undergraduate and postgraduate training in London (London School of Hygiene and Tropical Medicine, and St Bartholomew's Hospital Medical College) and his postdoctoral training at Tufts University School of Medicine and the University of Glasgow. Dr. Khan's primary research interests are in viral pathology and infectious diseases, with a particular focus on Epstein-Barr virus and its associated diseases. His secondary interest is in emerging viral infections and public health. He has received several awards for his teaching and research and is also a member of several national and international committees and organizations. He also serves on the editorial board of several international journals.

Abbreviations

APS	Ammonium persulfate
BGG	Bovine gamma globulin
BSA	Bovine serum albumin
cDNA	Complementary DNA
CHAPS	3-cholamidopropyl dimethylammonio-1-propanesulfonate
ChIP	Chromatin immunoprecipitation
COVID-19	Coronavirus disease-2019
CPE	Cytopathic effect
CSF	Cerebral spinal fluid
DAB	Diaminobenzidine
DAPI	4′,6-diamidino-2-phenylindole
DNA	Deoxyribonucleic acid
dNTP	Deoxyribonucleotide triphosphate
EBER	EBV-encoded RNA
EBV	Epstein–Barr virus
EDTA	Ethylenediaminetetraacetic acid
ELISA	Enzyme-linked immunosorbent assay
FFPE	Formalin-fixed, paraffin-embedded
FITC	Fluorescein isothiocyanate
HEPES	4-(2-hydroxyethyl)-1-piperazineethanesulfonic acid
HIV	Human immunodeficiency virus
HPLC	High-performance liquid chromatography
HSV	Herpes simplex virus
ICC	Immunocytochemistry
ICTV	International Committee on Taxonomy of Viruses
IHC	Immunohistochemistry
IP	Immunoprecipitation
ISH	In situ hybridisation
$MgCl_2$	Magnesium chloride
miRNA	microRNAs
mRNA	Messenger RNA
NBT/BCIP	p-nitroblue tetrazolium chloride/5-bromo-4-chloro-3-indolyl phosphate

NP-40	Nonyl phenoxypolyethoxylethanol-40
PAGE	Polyacrylamide gel electrophoresis
PBS	Phosphate-buffered saline
PCR	Polymerase chain reaction
qPCR	Quantitative PCR
RIPA	Radioimmunoprecipitation assay
RNA	Ribonucleic acid
RSV	Respiratory syncytial virus
RT-PCR	Reverse transcriptase PCR
SARS-CoV-2	Severe acute respiratory syndrome coronavirus-2
SCID	Severe combined immunodeficiency
SDS	Sodium dodecyl sulfate
siRNA	Small interfering RNA
SSC	Saline-sodium citrate
TBS	Tris-buffered saline
TEMED	Tetramethylethylenediamine
TMV	Tobacco mosaic virus

List of Figures

List of Tables

Introduction 1

1.1 What Are Viruses

Virology is the special branch of microbiology devoted to the study of viruses. Viruses are obligate intracellular parasites that infect both prokaryotic and eukaryotic cells. Outside their host cell, viruses exist as inert crystal particles (Crawford 2011) but remain infectious. The study of viruses started in the nineteenth century, with tobacco mosaic virus (TMV) being the first virus to be crystalised and characterised in 1935 (Pennazio 2010; Chaitanya 2019; Fauquet 2008). Viruses lack intracellular organelles essential for life, such as mitochondria, ribosomes, and replication machinery. Therefore, they cannot survive independently but depend on their host machinery for their basic life processes. For example, viruses use their host cell-generated nucleotides for replication and transcription and host ribosomes to translate their mRNA into proteins. They also lack an energy generation system; thus, they utilise energy from the host mitochondria for these processes. Viruses cannot generate or store ATP (Chaitanya 2019).

Viruses are made up of two primary components: genome and capsid, and in some cases, they may have an envelope (Prasad and Schmid 2012). Viral genomes can be either DNA or RNA, but unlike in eukaryotic cells, viruses do not contain both DNA and RNA (Crawford 2011). Although viruses are less complex than most other human pathogens, they account for a large proportion of human diseases, ranging from self-limiting infections such as the common cold to highly fatal infections such as rabies, Ebola, and even cancer (Khan et al. 2020; Lövheim et al. 2015; Drexler 2010). Over 200 viruses are known to infect humans (Woolhouse et al. 2012). They vary considerably in size, shape, genomic structure, replication strategy, and mode of transmission.

Z. Hassan, G. Khan, *Molecular Techniques for Studying Viruses*,
https://doi.org/10.1007/978-981-99-8097-0_1

1.2 Organisation of a Virus

A virus is a macromolecule consisting of nucleic acid (DNA or RNA) encapsulated in a proteinaceous capsid. There are proteins that are associated with the viral genome (nucleic acid); these are termed 'nucleoproteins'. The genome and the capsid together are called a 'nucleocapsid'. The nucleocapsid may or may not be surrounded by an envelope. A viral capsid is a coat made of single or multiple proteins. These proteins are self-assembled to give the capsid its characteristic structure. In addition to giving the virus its shape, the capsid protects the viral genome. Most viral capsids assume either helical or icosahedral symmetrical shapes. Some viruses contain an additional layer that covers the nucleocapsid and are thus referred to as enveloped viruses. Nucleocapsids without an envelope are referred to as nonenveloped viruses.

The viral envelope is derived from the host's plasma membrane; hence, it comprises a lipid bilayer. This envelope layer is decorated with viral glycoproteins, which determine the viral cell tropism. The viral genome encodes all the information needed to make a copy of the virus. Therefore, once inside a host cell, the virus uses the host cell organelles to make copies of itself. However, viruses differ in their degree of dependence on the host cell for replication. There are viruses that carry a limited number of viral-specific proteins that are not available in the host cell but are essential for the viral life cycle. These viral proteins include enzymes, such as protease and reverse transcriptase (Fig. 1.1).

1.3 Classification of Viruses

Although the basic structure of viruses is very simple; nucleic acid wrapped up in a protein shell, they differ considerably in their characteristics, such as size and shape, pathogenicity, host/tissue tropism, and the nature of their nucleic acid. These

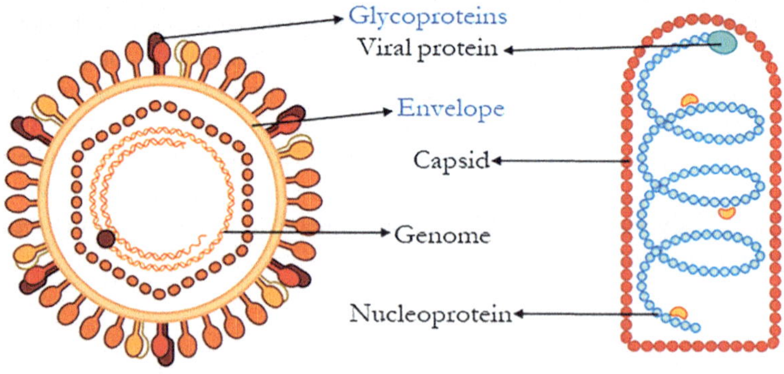

Fig. 1.1 Structural organisation of a virus. Components common to all viruses are indicated in black text, while those that can differ among viruses are shown in blue text. (Image created in BioRender)

differences formed the foundation for classifying and grouping viruses. In basic terms, viruses can be grouped into four types based on their nucleic acid:

- Double-stranded (ds) DNA viruses.
- Single-stranded (ss) DNA viruses.
- Double-stranded (ds) RNA viruses.
- Single-stranded (ss) RNA viruses.

These basic groups can be further subdivided, based on the presence or absence of an envelope, the nature of viral capsid symmetry (helical or icosahedral), and genome complexity. The viral genome can be segmented, linear or circular, positive sense (+), negative sense (−) or ambisense (Mahmoudabadi and Phillips 2018).

Initially, viruses were classified based on their phenotypic characteristics, such as organ tropism. For example, viruses infecting the liver were called hepatitis viruses; those infecting the gastrointestinal tract were called enteroviruses, and those infecting bacteria were called bacteriophages. This form of classification was very confusing and virtually impossible to implement for all viruses. For instance, at least five different viruses (hepatitis virus A to E) cause hepatitis. However, these viruses belong to different virus families and share very little characteristics other than their tropism for the liver. Some are DNA viruses, others RNA; some are enveloped, while others are nonenveloped.

The advent of the electron microscope led the International Committee on Taxonomy of Viruses (ICTV) to formulate a classification system based on morphological characteristics, mainly size and shape (Fauquet 2008; Mahmoudabadi and Phillips 2018). However, this system also failed to apply to all viruses. Therefore, in 1971, a method of classification based on the mechanism of mRNA synthesis was proposed. This system was established on the fact that all viruses need to synthesise mRNA to produce their proteins. This universal characteristic was central to the classification system, referred to as the Baltimore classification system. This system of classification provided, for the first time, a comprehensive method of classifying viruses based on their replication strategy (Baltimore 1971). Based on the Baltimore classification, viruses are grouped into seven groups. These are:

1. dsDNA → mRNA, e.g., Epstein–Barr virus, herpes simplex virus, papillomaviruses, poxviruses,
2. dsDNA → (+) ssRNA → dsDNA → mRNA, e.g., hepatitis B virus,
3. (+) ssDNA → dsDNA → mRNA, e.g., parvovirus
4. dsRNA → mRNA, e.g., rotavirus,
5. (+) ssRNA → (−) ssDNA → dsDNA → mRNA, e.g., HIV
6. (+) ssRNA → (−) ssRNA → mRNA, e.g., hepatitis C virus, poliovirus, coronaviruses
7. (−) ssRNA → mRNA, e.g., influenza virus, measles virus, rabies virus.

Since viruses differ considerably in their characteristics, studying them also involves several techniques. Therefore, the choice of a technique is crucial, and it should be based on the viral characteristics of interest.

1.4 How Are Viruses Studied

Since viruses are obligate intracellular parasites, they are often studied along with their host cell. There are several methods to study viruses. These include collecting virus-infected tissue and staining it for the presence and extent of infection (histology). Some viruses produce specific cellular changes such as cytopathic effect (CPE) in the infected tissues. For example, the rabies virus produces Negri bodies (eosinophilic, inclusion bodies) clearly visible in the cytoplasm of infected cells. Therefore, Negri bodies are characteristic features of rabies. Respiratory syncytial virus (RSV) can lead to the formation of syncytia (fusion of multiple cells). Additionally, many viruses are often released in body fluids, such as blood, urine, and cerebral spinal fluid (CFS), and they can be studied using molecular and cellular methods (Khan et al. 1991; Kanjilal et al. 2019). For example, techniques such as PCR and ELISA are used to detect viral nucleic acids and proteins, respectively.

Cell culture is used to isolate viruses from clinical samples. However, with the advent of molecular methods, the use of cell culture for routine diagnosis has declined considerably. Compared to cell culture, molecular techniques like PCR are not only just highly sensitive, quantitative, and rapid but also much safer. In addition, culturing highly pathogenic viruses such as Ebola, Lassa, and Crimean Congo haemorrhagic fever virus require specialised biosafety level 4 laboratory (Marx 2014). Thus, understanding the basic concepts governing the techniques in studying viruses is essential in selecting what is appropriate and plausible.

This book will focus on the laboratory procedures used in studying viruses at the molecular level—molecular virology. Molecular virology deals with the organisation and function of viral DNA/RNA and protein and the interactions between these macromolecules (Cann 2012). Molecular virology often overlaps with cell biology since these viruses and their components are studied with their host cell.

1.5 Basic Organisation of a Cell

A cell is a fundamental structural and functional unit of life. It is composed of and organised into many internal structures called organelles. The organelles perform specialised functions in the cell. For example, the ribosome is an organelle concerned with translating mRNA into proteins, and the mitochondria are concerned with the generation and storage of energy. Golgi sorts proteins to their destination organelle; for instance, proteins destined for degradation are sorted to the lysosome or to the plasma membrane for exocytosis.

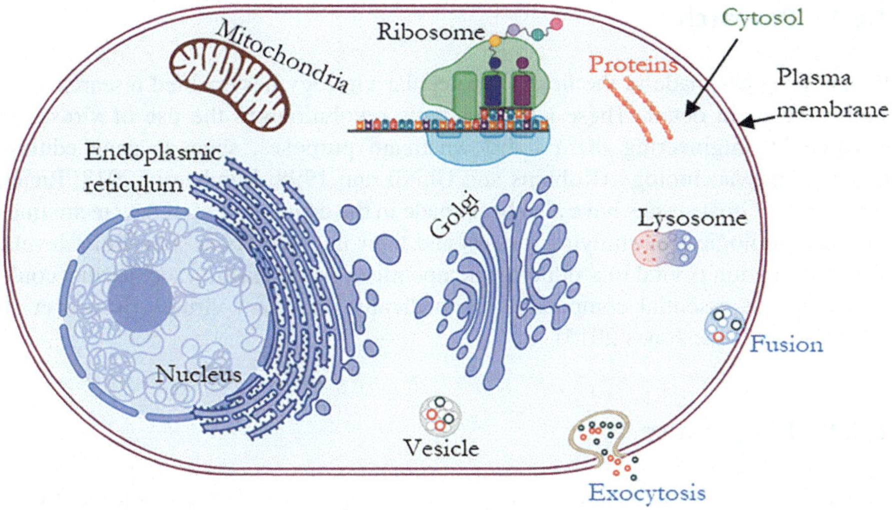

Fig. 1.2 Organisation of a eukaryotic cell. Membrane-bound organelles are indicated in black text, macromolecules in red text, cytosol in green text, and processes in blue text. (Image created in BioRender)

The inner fluid portion of the cell that contains all the organelles is called the cytosol. Cellular organelles are separated from one another by membrane structures (Katayama et al. 2020; Murat et al. 2010). The plasma membrane is the outer membrane that separates the internal organelles from the cell's surroundings. Cell membranes are semi-permeable; they allow the entry and exit of molecules/ substances. The main difference between a prokaryotic and eukaryotic cell is that the latter has a nucleus. The nucleus is a special organelle where DNA is stored, replicated, and transcribed (Fig. 1.2).

1.6 Application of Virology

Studying viruses is essential to understand and appropriately contain the fast-growing number of human viruses. Unlike other microorganisms such as *Saccharomyces cerevisiae* (baker's yeast) and probiotic bacteria (Hassan et al. 2016; Moyad 2008), viruses generally are not known to be beneficial to humans. However, it is possible that some viruses may indirectly benefit humans. For example, some viruses that asymptomatically infect humans could play a role in neutralising or preventing infection by more deadly pathogens in a symbiotic relationship (Mietzsch and Agbandje-McKenna 2017). Moreover, recent advances in molecular biology have led to the manipulation of viruses and adapting them as vectors for transferring genes (gene therapy). This approach has been used for treating a number of congenital diseases, including severe combined immunodeficiency (SCID) (Mietzsch and Agbandje-McKenna 2017; Robbins and Ghivizzani 1998).

1.6.1 Research

Recent progress made in the field of molecular virology has enabled researchers to study viruses in detail. These advances have revolutionised the use of viruses in research by engineering them for downstream purposes, such as gene editing, therapy, and vaccinology (Robbins and Ghivizzani 1998; Lundstrom 2018; Rauch et al. 2018). Great strides have also been made in the development of new techniques and methodologies for studying viruses and their interactions at subcellular levels. This information is vital in producing therapeutic or prophylactic vaccines that could target various essential components of medically important viruses (Rauch et al. 2018; Giese 2016; Jones 2015).

1.6.2 Diagnostics

Molecular viral techniques are widely used in diagnostic laboratories to detect the causative agent of diseases. For example, SARS-CoV-2, the causative agent of COVID-19, was identified using PCR (Gorbalenya et al. 2020; Zhu et al. 2020). Histopathological techniques, such as immunohistochemistry and in situ hybridisation, are used to detect and assess the site and extent of infections. Additionally, molecular pathology techniques can be used to visualise the pathological effects of a virus and its target tissue/cell (Duraiyan et al. 2012; Halling and Wendel 2008; McNicol and Farquharson 1997).

1.7 Scope of This Book

This book contains eleven chapters, including an introduction, nine chapters on molecular techniques used for studying viruses, and finally, a chapter on summary and conclusion. Each chapter focuses on outlining the general principles of a technique and the main practical steps involved. Possible problems that may be encountered during the experimental procedure and how to resolve them are provided at the end of each chapter. The objective is to provide an overview of the significant practical steps involved in each procedure. Thus, the book is intended to be an initial source for students starting their basic university laboratory training. The following techniques are covered: isolation of nucleic acids, isolation of proteins, PCR, western blotting, ELISA, immunoprecipitation, small interfering RNA, histological techniques, and bioinformatics and in silico stimulations. Relevant references are listed at the end of each chapter for readers requiring detailed protocols.

References

Baltimore D (1971) Expression of animal virus genomes. Bacteriol Rev 35:7
Cann A (2012) Principles of molecular virology. Academic Press, Amsterdam

Chaitanya KV (2019) Structure and organization of virus genomes. In: Genome and genomics, pp 1–30. https://doi.org/10.1007/978-981-15-0702-1_1
Crawford DH (2011) Viruses: a very short introduction. OUP Oxford, New York, NY
Drexler M (2010) Disease threats. what you need to know about infectious disease. National Academies Press, Washington, DC
Duraiyan J, Govindarajan R, Kaliyappan K, Palanisamy M (2012) Applications of immunohistochemistry. J Pharm Bioallied Sci 4:S307–S309
Fauquet CM (2008) Taxonomy, classification and nomenclature of viruses. Encyclopedia Virol:9–23. https://doi.org/10.1016/B978-012374410-4.00509-4
Giese M (2016) Introduction to molecular vaccinology. Springer International Publishing, Cham. https://doi.org/10.1007/978-3-319-25832-4
Gorbalenya AE et al (2020) Severe acute respiratory syndrome-related coronavirus: the species and its viruses—a statement of the Coronavirus Study Group. BioRxiv. https://doi.org/10.1101/2020.02.07.937862
Halling KC, Wendel AJ (2008) In situ hybridization: principles and applications for pulmonary medicine. In: Zander DS et al (eds) Molecular pathology of lung diseases. Springer, New York, NY, pp 117–129. https://doi.org/10.1007/978-0-387-72430-0_12
Hassan Z, Mustafa S, Rahim RA, Isa NM (2016) Anti-breast cancer effects of live, heat-killed and cytoplasmic fractions of Enterococcus faecalis and Staphylococcus hominis isolated from human breast milk. In Vitro Cell Dev Biol Anim 52:337–348
Jones LH (2015) Recent advances in the molecular design of synthetic vaccines. Nat Chem 7:952–960
Kanjilal S, Cho TA, Piantadosi A (2019) Diagnostic testing in central nervous system infection. Semin Neurol 39:297–311
Katayama T et al (2020) Isolation of a member of the candidate phylum 'Atribacteria' reveals a unique cell membrane structure. Nat Commun 11:6381
Khan G, Kangro HO, Coates PJ, Heath RB (1991) Inhibitory effects of urine on the polymerase chain reaction for cytomegalovirus DNA. J Clin Pathol 44:360–365
Khan G, Fitzmaurice C, Naghavi M, Ahmed LA (2020) Global and regional incidence, mortality and disability-adjusted life-years for Epstein-Barr virus-attributable malignancies, 1990-2017. BMJ Open 10:e037505
Lövheim H, Gilthorpe J, Adolfsson R, Nilsson L-G, Elgh F (2015) Reactivated herpes simplex infection increases the risk of Alzheimer's disease. Alzheimers Dement 11:593–599
Lundstrom K (2018) Viral vectors in gene therapy. Diseases 6:42
Mahmoudabadi G, Phillips R (2018) A comprehensive and quantitative exploration of thousands of viral genomes. elife 19:e31955
Marx V (2014) High-security labs: life in the danger zone. Nature 505:437–441
McNicol AM, Farquharson MA (1997) In situ hybridization and its diagnostic applications in pathology. J Pathol 182:250–261
Mietzsch M, Agbandje-McKenna M (2017) The good that viruses do. Ann Rev Virol 4:iii–v
Moyad MA (2008) Brewer's/baker's yeast (Saccharomyces cerevisiae) and preventive medicine: part II. Urol Nurs 28:73–75
Murat D, Byrne M, Komeili A (2010) Cell biology of prokaryotic organelles. Cold Spring Harb Perspect Biol 2:a000422
Pennazio S (2010) A history of virus diseases of plants: from the beginning to 1950. Riv Biol 103: 209–235
Prasad BVV, Schmid MF (2012) Principles of virus structural organization. Adv Exp Med Biol 726:17
Rauch S, Jasny E, Schmidt KE, Petsch B (2018) New vaccine technologies to combat outbreak situations. Front Immunol 9:1963

Robbins PD, Ghivizzani SC (1998) Viral vectors for gene therapy. Pharmacol Ther 80:35–47

Woolhouse M, Scott F, Hudson Z, Howey R, Chase-Topping M (2012) Human viruses: discovery and emergence. Philos Trans R Soc Lond Ser B Biol Sci 367:2864–2871

Zhu N et al (2020) A novel coronavirus from patients with pneumonia in China, 2019. N Engl J Med 382:727–733

2 Isolation of Nucleic Acids

Nucleic acid isolation is the process of extracting DNA or RNA from cells in a relatively pure form that can be used for downstream investigations, such as gene amplification and sequencing. Purified RNA must be converted to complementary DNA (cDNA) before using it for PCR. The reason for the conversion is that PCR works on the principle of a double-stranded macromolecule. Therefore, based on the downstream method and study objectives, nucleic acids can be isolated from either whole cellular or subcellular fractions (e.g., nucleus, cytoplasm, or membrane-bound organelles). This chapter will discuss the general principles of DNA/RNA isolation from the whole cell, nucleus, and cytoplasm. Additionally, converting RNA to cDNA is also covered as it is an essential prerequisite step for detecting RNA viruses using PCR.

2.1 Steps Involved in the Isolation of Nucleic Acids

There are three critical steps involved in extracting and purifying nucleic acids (DNA or RNA). These include disrupting the cell membrane, precipitating and enriching the molecule of interest, and finally resuspending the purified product in an appropriate buffer for downstream analysis.

1. **Cell disruption or cell lysis**: This step involves disrupting the cell membranes (most importantly, the plasma and nuclear membranes) to release the cellular contents into the solution. This disruption can be done by three mechanisms:
   (a) *Mechanical disruption*: includes homogenisation, sonication, or grinding. These methods physically break the membranes.
   (b) *Enzymatic lysis*: e.g., buffers that contain lysozyme or alkaline phosphatase. These methods enzymatically cleave the bonds that hold the membranes.
   (c) *Chemical methods*: e.g., lithium chloride, SDS, TRizol, triton X. These detergents chemically break down cell membranes (Hong et al. 1992). The

Z. Hassan, G. Khan, *Molecular Techniques for Studying Viruses*,
https://doi.org/10.1007/978-981-99-8097-0_2

breakdown happens due to the amphipathic nature of the membranes (i.e., having both hydrophilic and hydrophobic regions). Thus, it allows for the exchange of ions with the detergents, subsequently destroying the protein–protein bonds that hold the membrane.

2. **Precipitation and enrichment**: These are a series of steps where DNA or RNA is separated from the rest of the cellular components in the lysate.
   (a) Add phenol/chloroform to the cell lysate (obtained in step 1) to remove the unwanted cellular components such as proteins from the DNA (phenol/chloroform denatures and dissolves proteins).
   (b) Add RNase to degrade all RNA when DNA is needed and vice versa (add DNase when isolating RNA).
   (c) Add salts such as ammonium acetate (Singh et al. 2018) or lithium chloride (Kondo et al. 1991; Walker and Lorsch 2013). These salts interrupt the hydrogen bonds between the water and DNA molecules and make debris such as broken proteins and lipids clump and settle while DNA remains suspended.
   (d) Centrifuge at high speed to pellet down debris and clumps and collect the supernatant containing the nucleic acid of interest.
   (e) Wash and precipitate the nucleic acid of interest in cold alcohol (isopropanol, ethanol, or sodium acetate). Alcohol makes the nucleic acid aggregate and also removes salts. Nucleic acids are insoluble in alcohols. It is recommended that lithium chloride is used when high-quality RNA is needed, especially from small materials (Walker and Lorsch 2013).
   (f) Centrifuge at high speed to pellet down the nucleic acid of interest.
   (g) Discard the supernatant.
   (h) Additional wash in 70% ethanol solution is recommended to completely remove salts and other water-soluble impurities.
   (i) Centrifuge the nucleic acid of interest, discard the supernatant, and air-dry the pellet.
3. **Resuspension**: Resuspend the purified DNA or RNA in ultra-pure water or Tris-EDTA buffer to ensure stability and long-term storage. Resuspended RNA may require dissolving at 55–60 °C for 5–15 min.

Nucleic acid is now ready for downstream processes. It can be used immediately or stored at 4 °C for short-term or at −20 °C to −80 °C for long-term.

2.2 Materials

Centrifuge, heating block or water bath, mechanical disruptor, lysis buffer, DNase or RNase enzymes, alcohol, salt, ultra-pure water or Tris-EDTA, cells or tissue sample.

2.3 Isolation of DNA/RNA

As mentioned, DNA and RNA can be isolated from several cellular fractions. This chapter will discuss the major steps to consider while isolating RNA from the whole cell, nucleus, and cytoplasm. Additionally, it will highlight the general principles of DNA/RNA isolation from FFPE tissues. Troubleshooting tips are also highlighted (Table 2.1).

2.3.1 Isolation of DNA from Cells/Fresh Tissues

- Wash the cells or tissues twice in PBS.
- Follow the three general steps outlined in Sect. 2.1. During DNA precipitation, add RNase enzyme. Do not add DNase. (**Note**: Adding DNase will degrade DNA in the sample and result in no or poor yield.)
- Measure DNA concentration at 260 nm wavelength (Desjardins and Conklin 2010).

2.3.2 Isolation of RNA from Cells/Fresh Tissues

- Wash the cells or tissues twice in PBS.
- Follow the three general steps outlined in Sect. 2.1. During RNA precipitation, add DNase enzyme. Do not add RNase. (**Note**: Adding RNase will degrade RNA in the sample and result in no or poor yield.)
- Air-dry the RNA pellet and resuspend it in an appropriate buffer.
- Dissolve the resuspended RNA by heating at 55–60 °C for 5–15 min.
- Measure RNA concentration at 260 nm wavelength (Desjardins and Conklin 2010).

2.3.3 Isolation of DNA/RNA from Formalin-Fixed and Paraffin-Embedded Tissues

Tissues from routine biopsies or post-mortems are often stored in glass jars in formalin or in the form of formalin-fixed and paraffin-embedded (FFPE) blocks (Fig. 2.1). These FFPE blocks are regularly used for both clinical and research purposes. **Note**: Isolating RNA from FFPE sample is practically difficult. Extracting DNA from FFPE samples requires the following steps:

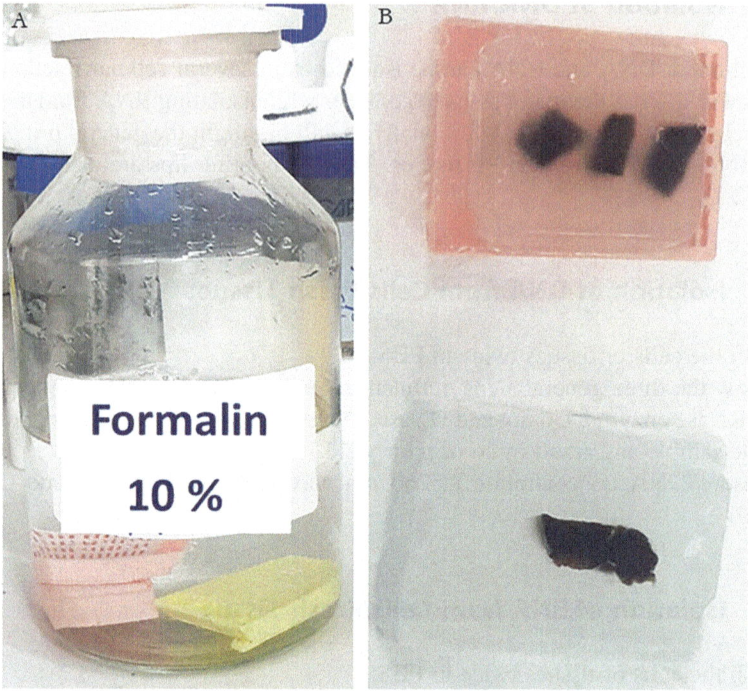

Fig. 2.1 Tissue in formalin and FFPE block. (**a**) Tissues in 10% formalin. (**b**) FFPE tissue blocks

- After sectioning the FFPE block (see Chap. 7 on histological methods, under 'sectioning'), pick 2–5 sections of ~10 μm into a tube.
- Pretreat the sections in xylene to dissolve the paraffin (Pikor et al. 2011).
- Wash the sections in ethanol and PBS (Pikor et al. 2011; Hassani and Khan 2015).
- Process the tissue sample in reverse order of the tissue processing steps (see Chap. 7 on 'histological methods, graded ethanol').
- Digest the pretreated tissue in proteinase K buffer.
- Follow the three steps in DNA/RNA isolation (outlined in Sect. 2.1—membrane disruption, nucleic acid precipitation and enrichment, and then resuspension).
- Measure DNA concentration at 260 nm wavelength (Desjardins and Conklin 2010).

Since formalin fixation causes the crosslinking of proteins and nucleic acids, it often randomly breaks nucleotide sequences (Hassani and Khan 2015; Sarnecka et al. 2019). Thus, the DNA extracted from FFPE is fragmented and of poor quality, making it difficult to be amplified (Hassani and Khan 2015). (**Note**: It is crucial to digest the tissue in proteinase K buffer to eliminate proteins and enzymes that may interfere with downstream processes. (Pikor et al. 2011; Hassani and Khan 2015)).

2.3.4 Isolation of Subcellular DNA/RNA

Subcellular fractions (nucleus and cytoplasm) can be purified for isolation of DNA/RNA from these fractions (Greenberg and Bender 2007). Isolation of DNA from the cytoplasm is particularly important for the detection of cytoplasmic DNA viruses such as poxviruses or other DNA viruses that assemble and egress through the cytoplasm (Hassan et al. 2021; Shimizu et al. 2014).

- Centrifuge the cell suspension and wash twice in PBS. (**Note**: Centrifugation at low speed is important to avoid cells' rupture, which may interfere with the cytoplasmic yield.)
- Lyse the cells in weak buffers or detergents such as RLN buffer or nonyl phenoxypolyethoxylethanol-40 (NP-40) for 2 min at 4 °C. (**Note**: Using strong lysis reagents at this point may also disintegrate the nuclear membrane, thereby mixing the two subcellular fractions.)
- Centrifuge at low speed for 2 min at 4 °C to pellet the nuclei.
- Wash the nuclei pellet twice in PBS.
- Subject the supernatant to differential centrifugation at 4 °C at 3000 × g for 5 min, followed by 12,000 × g for 5 min, and finally, at 15,000 × g for 5 min.

Now the nuclear and cytoplasmic fractions are ready. Perform DNA or RNA isolation as described above (in Sect. 2.1—i.e., membrane disruption, nucleic acid precipitation and enrichment, and then resuspension). Determine the purity and concentration of the subcellular samples. Then, perform PCR to check for the efficiency of the subcellular fractionation by amplifying spliced and unspliced housekeeping genes, such as β-actin (see Chap. 4 under 'types of PCR'). Separate the amplicon on an agarose gel (see Chap. 4 under 'visualisation of PCR amplicon'). **Note**: Spliced β-actin is expected to be amplified in both nuclear and cytoplasmic fractions, whereas the unspliced β-actin should be only in the nuclear fraction. This is because only spliced and mature mRNA is transported to the cytoplasm (Ohno et al. 2002; Hilleren et al. 2001).

2.4 Determining the Quality and Concentration of Purified Nucleic Acids

The purity and concentrations of DNA and RNA are determined by nanoanalysers such as nanodrop or spectrophotometers (Fig. 2.2). The concentration of DNA or RNA is measured at 260 nm wavelength. The sample purity is indicated by the ratio of absorbance at 260 nm and 280 nm. Generally, a ratio of 1.8–2.0 is considered a pure sample. Alternatively, the amount of DNA or RNA in the purified sample can

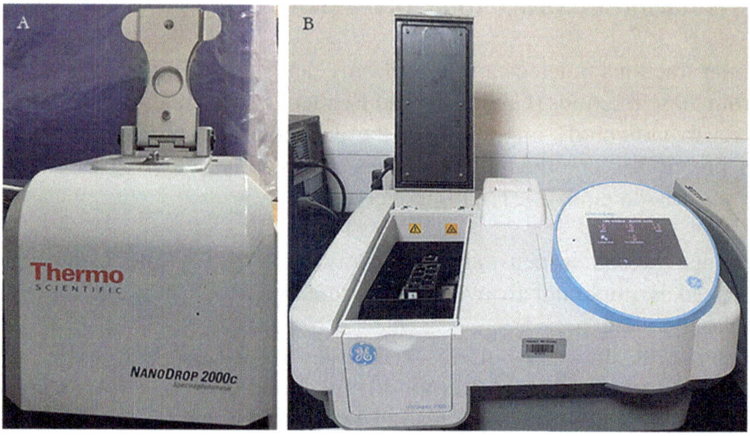

Fig. 2.2 Instruments for quantification of macromolecules. (**a**) Nanodrop. (**b**) Spectrophotometer

be determined using PCR (see Chap. 4 on 'PCR'), followed by separating and visualising the products on agarose gel (see Chap. 4 on 'agarose gel electrophoresis'). (**Note**: RNA must first be reverse transcribed to cDNA before PCR.) The intensity of the DNA bands on the gel semi-quantitates the concentration of DNA in that sample. This semi-quantification can be used for a relative comparison of DNA or RNA between samples or the viral load.

2.5 Reverse Transcription

Purified RNA is reverse transcribed to its complementary DNA using the enzyme, reverse transcriptase. The process of reverse transcription can be done either before or during PCR.

- To an appropriate concentration of RNA (1–5 μg), add a reverse transcriptase enzyme and a primer. The primer can be generic or gene-specific. (**Note**: If gene-specific primers are used during the reverse transcription, only the specific gene will be reverse transcribed.)
- Incubate the reaction at the optimum working temperature of the reverse transcriptase (37–42 °C) for 30–60 min.
- Inactivate the enzymes at a high temperature (90–95 °C) for 5 min. (**Note**: Keeping the reaction for long at high temperature may denature and damage the cDNA.)
- Centrifuge to collect all vapours in the tube lid. The cDNA is now ready. It can be used immediately or store at 4 °C for short-term or − 20 °C to −80 °C for short-term.

2.6 Troubleshooting

Table 2.1 Troubleshooting for nucleic acids isolation

S/ No	Problem	Possible reason(s)	Solution(s)
1.	No DNA/RNA in the purified sample	• Insufficient starting material • Mishandling of sample • Improper proteinase K digestion of FFPE samples	• Check the quality of starting materials • Optimise precipitation • Ensure that all samples are at the right temperature during and after isolation • Add low SDS concentration in proteinase K buffer • Precipitate the DNA with a buffer containing low NaCl concentration
2.	No cDNA in the reverse transcribed sample	• RNA is degraded • Lack of enzyme-buffer compatibility	• Make sure the RNA is present and of good quality • Check the compatibility of the components in the reverse transcription reaction mixture
3.	Impurities in DNA/ RNA sample	• Impurities were not properly precipitated • The nucleic acid of interest is not properly enriched	• Digest impurities by using RNase (for DNA isolation) or DNase (for RNA isolation) • Optimise precipitation • Increase washing steps
4.	Mixed up of nuclear and cytoplasmic nucleic acids	• High centrifugation speed during cell fractionation • Use of strong cell lysis reagents, e.g., TRizol, SDS, etc	• Reduce centrifugation speed • Use less strong cell lysis reagents

References

Desjardins P, Conklin D (2010) NanoDrop microvolume quantitation of nucleic acids. J Vis Exp 45:2565. https://doi.org/10.3791/2565

Greenberg ME, Bender TP (2007) Identification of newly transcribed RNA. Curr Protoc Mol Biol Chapter 4:Unit 4.10

Hassan Z, Kumar ND, Reggiori F, Khan G (2021) How viruses hijack and modify the secretory transport pathway. Cell 10:2535

Hassani A, Khan G (2015) A simple procedure for the extraction of DNA from long-term formalin-preserved brain tissues for the detection of EBV by PCR. Exp Mol Pathol 99:558–563

Hilleren P, McCarthy T, Rosbash M, Parker R, Jensen TH (2001) Quality control of mRNA 3′-end processing is linked to the nuclear exosome. Nature 413:538–542

Hong Y-K, Coury DA, Polne-Fuller M, Gibor A (1992) Lithium chloride extraction of DNA from the seaweed Porphyra Perforata (rhodophyta)1. J Phycol 28:717–720

Kondo T, Mukai M, Kondo Y (1991) Rapid isolation of plasmid DNA by LiCl-ethidium bromide treatment and gel filtration. Anal Biochem 198:30–35

Ohno M, Segref A, Kuersten S, Mattaj IW (2002) Identity elements used in export of mRNAs. Mol Cell 9:659–671

Pikor LA, Enfield KSS, Cameron H, Lam WL (2011) DNA extraction from paraffin embedded material for genetic and epigenetic analyses. J Vis Exp 49:2763. https://doi.org/10.3791/2763

Sarnecka AK et al (2019) DNA extraction from FFPE tissue samples—a comparison of three procedures. Contemp Oncol (Pozn) 23:52–58

Shimizu A et al (2014) Characterisation of cytoplasmic DNA complementary to non-retroviral RNA viruses in human cells. Sci Rep 4:5074

Singh UA, Kumari M, Iyengar S (2018) Method for improving the quality of genomic DNA obtained from minute quantities of tissue and blood samples using Chelex 100 resin. Biol Proced Online 20:12

Walker SE, Lorsch J (2013) Chapter nineteen-RNA purification–precipitation methods. In: Lorsch J (ed) Methods in enzymology, vol 530. Academic Press, pp 337–343

Isolation of Proteins

3

Proteins are fragile macromolecules that are highly sensitive to environmental conditions. Due to their chemical and/or biological instability, proteins can be degraded during experiment (Lee 2017). This loss is particularly a practical limitation with low abundantly expressed proteins. Therefore, it is important to obtain a relatively high protein concentration at the beginning of the experiment.

3.1 Steps Involved in the Isolation of Proteins

Since proteins are sensitive to environmental conditions, it is crucial to perform all isolation steps on ice to prevent degradation during these processes. There are several important steps in isolating proteins. These include disrupting the cell membrane, protein enrichment, and stabilisation.

1. **Cell disruption or cell lysis**: This step involves disrupting the cell membranes (most importantly, the plasma and nuclear membranes) to release the cellular contents into the solution. This disruption can be done by three mechanisms:
   (a) *Mechanical disruption*: includes homogenisation, sonication, or grinding. These methods physically break the membranes.
   (b) *Enzymatic lysis*: e.g., buffers that contain lysozyme or alkaline phosphatase. These methods enzymatically cleave the bonds that hold the membranes.
   (c) *Chemical methods*: e.g., radioimmunoprecipitation assay (RIPA) buffer, 3-cholamidopropyl dimethylammonio-1 propanesulfonate (CHAPS), or 4-(2-hydroxyethyl)-1-piperazineethanesulfonic acid (HEPES) buffer, SDS, TRizol, triton X. These detergents break down cell membranes (Hong et al. 1992) and solubilise proteins (Churchward et al. 2005).
2. **Precipitation and enrichment**: These are a series of steps where proteins are separated from the rest of the cellular components in the lysate.

Z. Hassan, G. Khan, *Molecular Techniques for Studying Viruses*,
https://doi.org/10.1007/978-981-99-8097-0_3

(a) Add protease/phosphatase inhibitors to the cell lysate (obtained in step 1) to prevent the degradation of proteins and maintain their posttranslational modifications such as phosphorylation.
(b) Add DNase and RNase to degrade nucleic acids (optional).
(c) Add reducing agents, such as β-mercaptoethanol to the lysate (Optional). This will denature RNases and unfold the native conformation of the protein, thereby making it more suitable for SDS-PAGE.
(d) Incubate the mixture at 4 °C for 10–15 min.
(e) Centrifuge at high speed to pellet down debris and clumps.
(f) Collect the supernatant containing the protein and discard the pellet.
(g) Process protein immediately for downstream experiment or store at 4 °C for short-term or at −20 °C to −80 °C for long-term.

Since proteins, especially those expressed at low abundance, have the tendency to get lost in a pool of a mixed sample, purification of these proteins may be necessary for their study. Individual proteins of known characteristics can be purified from the total proteins. For this, methods, such as column chromatography, separate protein mixtures based on their size, shape, or charge (Coskun 2016). Other chromatography techniques, including high-performance liquid chromatography (HPLC), purify proteins based on their affinity to form polar or dipolar interactions with stationary phases like silica. HPLC can be used to purify many macromolecules, including amino acids and peptides (Coskun 2016).

This chapter will focus on isolating a mixture of total proteins present in three cellular fractions. These are the whole cell, nucleus, and cytoplasm. Troubleshooting tips are highlighted (Table 3.1).

Table 3.1 Troubleshooting for protein isolation

S/N	Problem	Possible reason(s)	Solution(s)
1.	No protein in the sample	• Insufficient starting material • Mishandling of sample	• Check the quality of starting materials • Ensure that all samples are at a low temperature during and after isolation
2.	Mixing of nuclear and cytoplasmic proteins	• Use of strong lysis solution for cytosolic/membrane-bound protein • Longer incubation during cell lysis	• Optimise lysis buffer strength and incubation time
3.	High variability in protein concentration from the same sample	• Not the same protein volume is added to the protein assay reaction • Presence of impurities, e.g., detergent, surfactant, or antimicrobials that interfere with the downstream experiment	• Check pipetting skills • Optimise homogenisation for tissue samples • Ensure the suitability of the lysis method for the choice of the downstream experiment

3.2 Materials

Centrifuge, heating block or water bath, mechanical disruptor, lysis buffer, protease and phosphatase inhibitors, Dnase or Rnase enzymes, β-mercaptoethanol, cells or tissue sample.

3.3 Isolation of Proteins

Based on the downstream method and study objectives, proteins can be isolated from either whole cellular or subcellular fractions (e.g., nucleus, cytoplasm, or membrane-bound organelles). To obtain proteins from these fractions, specific lysis procedures are essential.

3.3.1 Isolation of Whole Cell Proteins

- To purify proteins from viral infected cells or tissues, treat the sample with buffers containing strong detergents such as RIPA, SDS, CHAPS, or HEPES buffers. In addition to disrupting the cell membrane by breaking its protein–protein interactions, these buffers are also known to solubilise proteins (Churchward et al. 2005). (**Note**: Cells can be lysed by resuspending cell pellets in these buffers, whereas tissues may require prior mechanical disruption of extracellular matrix to form cell suspension. This disruption can be done by either grinding or homogenisation.) Further, supplementing the lysis buffer with β-mercaptoethanol, Dnase and Rnase, and inhibitors of protease and phosphatase (e.g., phenylmethylsulfonyl fluoride) will help to stabilise proteins better and improve the overall protein integrity and purity.
- Incubate the mixture at 4 °C for lysis to occur. (**Note**: Maintaining 4 °C throughout the process of protein isolation is crucial to avoid protein degradation.)
- Centrifuge at 4 °C at high speed (13,000 × g) for 10 min to pellet down debris.
- Collect the supernatant as total cellular proteins.
- Proteins can be used immediately or store at 4 °C for short-term storage or at −80 °C for long-term storage.

3.3.2 Isolation of Subcellular Proteins

Isolating proteins from subcellular fractions requires ensuring that the membrane structures that separate the organelles remain intact until they are meant to be broken. Therefore, sequential lysis of these membranes is crucial for isolating subcellular proteins. Practically, the selection of lysis buffer for lysing each membrane is essential. (**Note**: Supplementing lysis buffer at each step with chemical intermediates that will selectively allow the lysis of a membrane and stabilise others is necessary.)

Hexylene glycol is a good chemical intermediate for membrane stabilisation (Baghirova et al. 2015; Holden and Horton 2009).

- To purify subcellular proteins from viral infected cells or tissues, treat the sample with weak detergents such as digitonin. (**Note**: Cells can be lysed by resuspending cell pellet in the buffer, whereas tissues may require prior gentle mechanical separation to disrupt its extracellular matrix and form cell suspension. The disruption can be done by either grinding or homogenisation in digitonin.)
- Incubate the mixture at 4 °C for 5–10 min to lyse the plasma membrane (**Note**: Incubation time may need to be optimised based on cells' nuclear to cytoplasmic ratio.)
- Centrifuge at 4 °C at low speed (2000–4000 × g) for 10 min.
- Collect supernatant as cytosolic proteins.
- Wash the pellet twice in PBS and resuspend in more strength buffers such as NP-40.
- Incubate the mixture at 4 °C for 10–30 min to lyse the membranes of internal organelles. (**Note**: Incubation time may need to be optimised based on cells' nuclear to cytoplasmic ratio.)
- Centrifuge at 4 °C at medium speed (5000–7000 × g) for 10 min.
- Collect supernatant as membrane-bound proteins.
- Wash the pellet twice in PBS and resuspend in more strong buffers such as SDS supplemented with DNase and RNase.
- Incubate the mixture at 4 °C for 10–30 min to lyse the nuclear membranes.
- Centrifuge at 4 °C at high speed (8000–13,000 × g) for 10 min to pellet down debris.
- Collect supernatant as nuclear proteins.

Note: Each lysis buffer is supplemented with hexylene glycol (for membrane stability), HEPES (for pH stability), sodium chloride (for ionic strength), and protease and phosphatase inhibitors to prevent the degradation of proteins.

3.4 Determination of Protein Purity and Concentration

As with nucleic acids, nanoanalysers or spectrophotometers are used to determine the purity and concentration of proteins. Additionally, there are several biochemical methods available. These assays work on reducing a substance to produce a colour change. They include BCA, Lowry, and Bradford (Coomassie) assays. A standard protein of known concentration is required to compare the test sample. Therefore, a highly pure protein and stable protein is usually used to generate a reliable calibration curve. For example, bovine serum albumin (BSA) and bovine gamma globulin (BGG) are commonly used as standards to determine the concentrations of proteins and antibodies, respectively.

The Beer-Lambert law (Rodger 2013; Beer–Lambert law 2014) is used to work out the concentration of a protein to be used as a standard for determining the concentration of an unknown sample.

$$\text{Beer} - \text{Lambert equation}, A = \varepsilon l c$$

where:

A = Absorbance.

ε = Molar extinction coefficient.

l = Length of solution the light travels through.

c = Concentration of a given solution.

The procedure for determining the concentration of an unknown protein sample from BSA standard by Bradford assay is outlined as below:

- Dilute a small amount of the protein sample (1–2 μL) in 1 mL of a buffer such as Tris-HCl with Coomassie Brilliant Blue protein-dye added in a specific ratio (e.g., 4:1 ratio). (**Note**: HCl is added to the buffer to create an acidic environment which allows proper binding of proteins to the dye.)
- Incubate the mixture at room temperature for 30–45 min for the reaction to occur. The reaction involves reducing Coomassie dye and producing a colour change from reddish-brown dye colour to blue. The extent of colour change is directly proportional to the amount of protein in the sample.
- Measure the absorbance using a spectrophotometer (Chap. 2, Fig. 2.2) at 595 (or 570–620) nm wavelength (Martz n.d.).
- Calculate the protein concentration against the concentration of a standard calibration curve. (**Note**: Experiments of standard and sample must be performed simultaneously to minimise errors.)

Table 3.2 shows the volume of 0.2 μg/μL BSA (40 μg BSA diluted in 200 μL water) along with the known protein quantities required to generate a calibration curve.

A standard calibration curve is generated by plotting absorbance obtained for the standards against its corresponding concentration (Martz n.d.). The slope and intercept of the standard calibration curve are then used to determine the concentration of the test protein samples. The linear correlation coefficient (Hong et al. 1992) expresses the accuracy of the standard curve (Fig. 3.1).

The concentration of unknown protein is calculated using the following formula:

$$\text{Protein concentration} = \frac{\text{A} - \text{I}}{\text{S}}$$

where:

A = Absorbance of the test sample.

I = Intercept of the standard curve on the y-axis.

S = Slope of the standard curve.

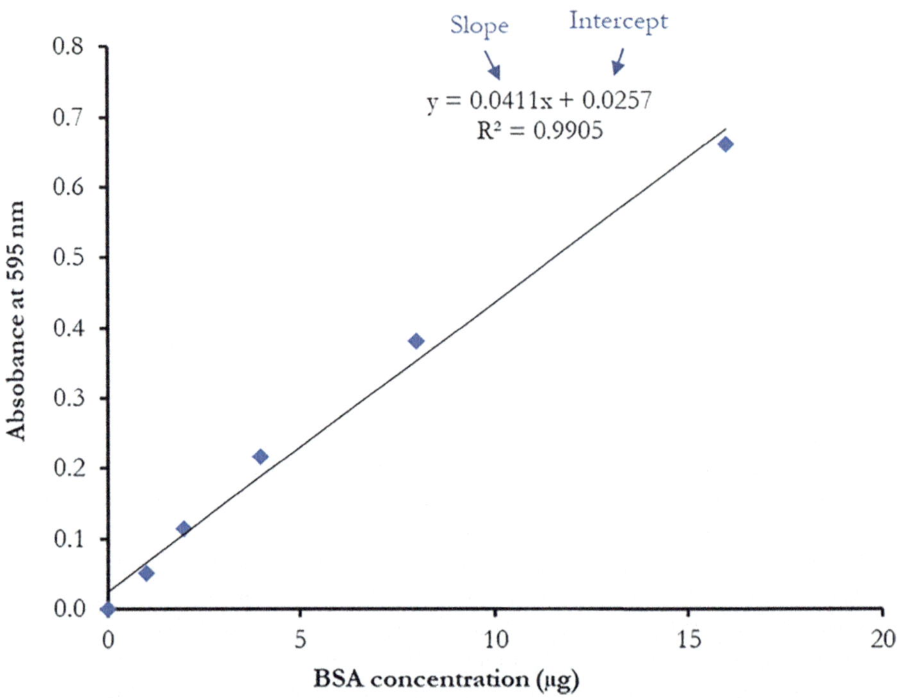

Fig. 3.1 BSA standard curve calibration graph. A BSA standard curve calibration graph showing the equation with slope, intercept, and R^2 of 99.05% accuracy

3.5 Troubleshooting

Table 3.2 Preparation of a standard calibration curve

S/N	Vol. of diluted BSA (µL)	50 mM Tris-HCl (µL)	Coomassie dye (µL)	Final quantity of BSA (µg)
1.	0	800	200	0
2.	5	795	200	1
3.	10	790	200	2
4.	20	780	200	4
5.	40	760	200	8
6.	80	720	200	16

References

Baghirova S, Hughes BG, Hendzel MJ, Schulz R (2015) Sequential fractionation and isolation of subcellular proteins from tissue or cultured cells. MethodsX 2:e440–e445

Beer–Lambert law (2014) Beer–Lambert law. In: Gilbert-Kawai ET, Wittenberg MD (eds) Essential equations for anaesthesia: key clinical concepts for the FRCA and EDA. Cambridge University Press, London, pp 43–44. https://doi.org/10.1017/CBO9781139565387.023

Churchward MA, Butt RH, Lang JC, Hsu KK, Coorssen JR (2005) Enhanced detergent extraction for analysis of membrane proteomes by two-dimensional gel electrophoresis. Proteome Sci 3:5

Coskun O (2016) Separation techniques: chromatography. North Clin Istanb 3:156–160

Holden P, Horton WA (2009) Crude subcellular fractionation of cultured mammalian cell lines. BMC Res Notes 2:243

Hong Y-K, Coury DA, Polne-Fuller M, Gibor A (1992) Lithium chloride extraction of DNA from the seaweed Porphyra Perforata (Rhodophyta)1. J Phycol 28:717–720

Lee CH (2017) A simple outline of methods for protein isolation and purification. Endocrinol Metab 32:18–22

Martz E. Bradford assay for protein. http://www.bio.umass.edu/micro/immunology/542igg/bradford.htm

Rodger A (2013) Beer-Lambert law derivation. In: Roberts GCK (ed) Encyclopedia of biophysics. Springer, Berlin, pp 184–185. https://doi.org/10.1007/978-3-642-16712-6_783

PCR-Based Techniques

4

Polymerase chain reaction (PCR), a molecular biology technique that amplifies and quantifies genes, was invented by Kary Mullis in 1983 and patented in 1985 (Saiki et al. 1985). This Nobel Prize-winning technique has revolutionised the detection and analysis of nucleic acids. It is essentially an in vitro enzymatic reaction for amplifying DNA using thermal cycling (Kadri 2019). The procedure involves an initial denaturation step, followed by 30–50 cycles of amplification, and then a final extension step (Kadri 2019).

- Initial denaturation at 92–95 °C for 5 min separates the two-helix double-strands of DNA and eliminates secondary structures.
- Amplification involves 30–50 cycles of *denaturation* at 92–95 °C for 30 s - 1 min, *primer annealing* (based on primer, most often 45–55 °C for 30–45 s), and *extension* at 72 °C for 60 s.
- Perform final extension at 72 °C (72 °C is the optimal temperature for *Taq* DNA polymerase activity) for 10 min.

At least one positive and negative control should be included in each PCR experiment. (**Note**: Primers anneal at ~5 °C below their melting temperature (T_m)).

$$T_m = 4(G + C) + 2(A + T)$$

where,

T_m = Melting temperature of the primer.
G = Guanine (nucleotide).
C = Cytosine (nucleotide).
A = Adenine (nucleotide).
T = Thymine (nucleotide).

Z. Hassan, G. Khan, *Molecular Techniques for Studying Viruses*,
https://doi.org/10.1007/978-981-99-8097-0_4

4.1 Types of PCR

There are different types of PCRs. However, the fundamental practical steps are the same for all. The difference lies in the nature of the starting material, i.e., the DNA template, and how the results are obtained. PCR results are obtained either at the end of the experiment (endpoint) or during the experiment (real-time).

This chapter will focus on the common types of PCR. They include:

4.1.1 Standard PCR

Standard PCR: Uses purified DNA as the template.

4.1.2 Reverse Transcriptase PCR

Reverse transcriptase PCR (RT-PCR): Uses synthesised cDNA that was reverse transcribed from a purified RNA as the template (Kadri 2019; Wacker and Godard 2005).

4.1.3 One-Step PCR

One-step PCR: Uses purified RNA that is converted to cDNA during the PCR cycling (Wacker and Godard 2005).

4.1.4 Quantitative PCR

Quantitative PCR (qPCR), also known as real-time PCR: Can be any of the above but generates numerical data in real-time. This numerical data is obtained from fluorescent signals emitted in proportion to the PCR amplified product after each cycle (Kadri 2019). Therefore, the thermal cycler for qPCR is coupled to an optical reading system.

4.1.5 Multiplex PCR

Multiplex PCR: Can be any of the above types of PCR but uses more than one set of primers in the reaction (Marmiroli and Maestri 2007; Shen 2019; Mahony and Chernesky 1995). A PCR with only one set of primers is called monoplex PCR. (**Note**: Melting temperature of the different primer sets for multiplex PCR must be compatible for a successful reaction. Therefore, primers must anneal and dissociate from their complementary DNA at the same temperature (Mahony and Chernesky 1995)). Multiplex PCR is particularly useful in screening for multiple viruses simultaneously in a single clinical sample (Wittwer et al. 2001). A typical example where multiplex PCR is preferred is in the diagnosis of respiratory tract infections. Respiratory tract infections are caused by many different viruses, often presenting with similar clinical features (Zumla et al. 2014).

4.1.6 Nested PCR

Nested PCR: This involves amplifying two monoplex PCRs separately. The first PCR uses a primer set (outer primer) that amplifies a larger portion of a gene. Products obtained from this PCR are then used as templates for the second PCR reaction. The second PCR uses a primer set (inner primers) that amplifies a region lying within the fragment amplified in the first round (Green and Sambrook 2019).

4.2 Materials

Thermocycler, PCR master mix (PCR mix, $MgCl_2$, dNTPs, Taq polymerase, Taq buffer), primers (forward and reverse), (**note**: forward primers amplify the leading strand of the DNA, while reverse primer amplifies the lagging strand), electrophoresis apparatus, Gel doc, and template.

4.3 Procedure

4.3.1 PCR Amplification

- Prepare PCR master mix, add primers (gene-specific or universal primers) and the template (DNA, cDNA, or RNA).
- Amplify the mixture in a thermocycler following the three PCR steps: denaturation, primer annealing, and extension.

4.3.2 Visualisation of PCR Amplicon

qPCR data can be visualised in real-time as the amplification progresses (Fig. 4.1 for qPCR amplification curve). Melt curve analysis is required when certain dyes, such as SYBR green, are used (Fig. 4.1 for qPCR melt curve plot). This analysis is required because SYBR green intercalates into any ds DNA in the reaction. These

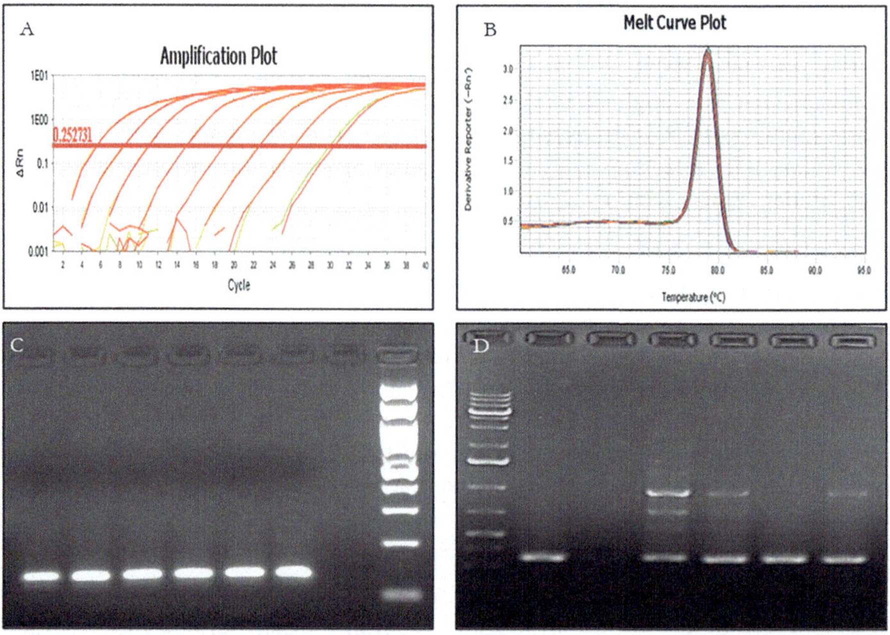

Fig. 4.1 PCR result visualisation formats. (**a**) Amplification plot for qPCR. (**b**) Product melt curve. (**c**) Agarose gel electrophoresis of end-point PCR products from a monoplex PCR. (**d**) Agarose gel electrophoresis of end-point PCR products from a multiplex PCR

include primer dimers or any nonspecific primer binding. A single peak in the melt curve plot indicates the presence of a single product in the reaction. Conversely, end-point PCR products need to be separated on an agarose gel (Kadri 2019). End-point data can be semi-quantitated by measuring the intensity of the bands (Fig. 4.1 for gel images).

4.3.3 Agarose Gel Electrophoresis

To perform agarose gel electrophoresis, prepare a gel of certain percentage. The gel percentage depends on the molecular weight of the PCR amplicon, i.e., low compact gels are required for high molecular weight amplicon and vice versa.

- Dissolve appropriate percentage, e.g., 1-2% (w/v) agarose in buffers such as Tris-acetate EDTA or Tris-borate EDTA by heating.

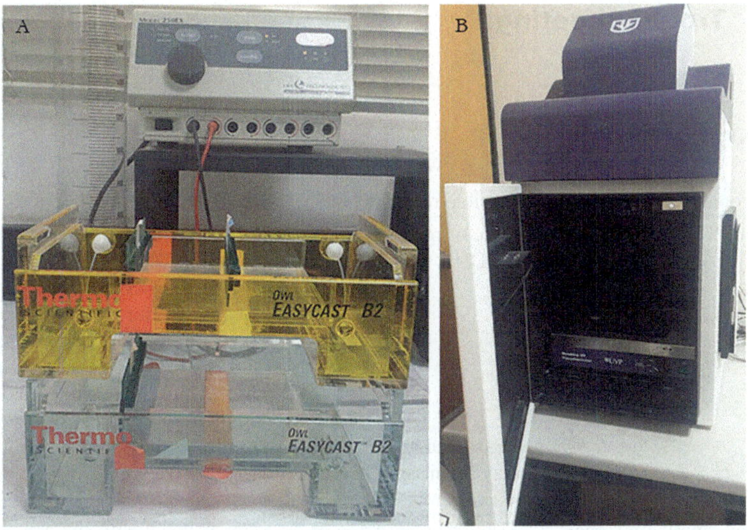

Fig. 4.2 Instruments for agarose gel electrophoresis. (**a**) Run set up for agarose gel electrophoresis. (**b**) UV trans-illuminator for visualisation of agarose gel

- Add a few drops of DNA intercalating dyes such as ethidium bromide or SYBR green.
- Pour the gel into a preassembled gel electrophoresis tray, insert a gel comb of desired thickness, and let it solidify (Fig. 4.2).
- Once solidified, remove the comb, and place the gel in a gel tank containing the same buffer used to prepare the gel.
- Load PCR products into the wells, along with a DNA ladder of known molecular weight.
- Run the electrophoresis at a voltage of 90–120 v for 30–60 min.
- View the gel under a UV trans-illuminator and capture the image (Fig. 4.2).

Troubleshooting tips are highlighted (Table 4.1).

4.4 Troubleshooting

Table 4.1 Troubleshooting for PCR

S/N	Problem	Possible reason(s)	Solution(s)
1.	No PCR amplicon	• No, or poor quality of the DNA template • Essential components may be missing in the reaction mixture	• Check the quality and concentration of starting DNA • Decrease SDS in proteinase K digestion buffer for FFPE samples • Avoid amplifying larger fragments (500 bp) from FFPE DNA samples • Make sure the reaction mixture is complete • Optimise the component mixture
2.	Low amplicon signal	• Low concentration of DNA template • Insufficient cycling number • Self-annealing of primers • Not equal proportion of PCR mix reagents	• Check the quality and concentration of starting DNA • Increase PCR cycles • Check the compatibility of forward and reverse primers • Optimise the amount of each component in the PCR mix
3.	Nonspecific amplification	• Nonspecific binding of primers • Self-annealing of primers • Possibility of variant (s) of the gene of interest	• Check primer specificity and optimise the primer annealing temperature • Do a hot-start PCR (heating the thermocycler until it reaches the denaturation temperature before loading the sample tubes) • Add primer inhibitors to the reaction mixture[a] • Check the compatibility of forward and reverse primers • Design primers to target a different region in the gene of interest
4.	Presence of primer dimers	• Excess amount of primer • Self-annealing of primers	• Reduce primer concentrations • Check the compatibility of forward and reverse primers
5.	Melt curve analysis showed double peaks	• Too many primers or primer dimers • Nonspecific primer binding	• Optimise primer concentrations • Check the compatibility of forward and reverse primers • Optimise annealing temperature
6.	Target with complex secondary structures		• Optimise denaturation temperature
7.	Low abundance/small product length		• Do a nested PCR

[a]Primer inhibitors are molecules that prevent primer binding. These inhibitors are often destroyed by heat; therefore, primers will be available for specific binding after the initial denaturation

References

Green MR, Sambrook J (2019) Nested polymerase chain reaction (PCR). Cold Spring Harb Protoc 2019:2

Kadri K (2019) Polymerase chain reaction (PCR): principle and applications. In: Synthetic biology—new interdisciplinary science. IntechOpen. https://doi.org/10.5772/intechopen.86491

Mahony JB, Chernesky MA (1995) 10—Multiplex polymerase chain reaction. In: Wiedbrauk DL, Farkas DH (eds) Molecular methods for virus detection. Academic Press, San Diego, pp 219–236. https://doi.org/10.1016/B978-012748920-9/50011-X

Marmiroli N, Maestri E (2007) Chapter 6—polymerase chain reaction (PCR). In: Picó Y (ed) Food toxicants analysis. Elsevier, pp 147–187. https://doi.org/10.1016/B978-044452843-8/50007-9

Saiki RK et al (1985) Enzymatic amplification of beta-globin genomic sequences and restriction site analysis for diagnosis of sickle cell anemia. Science 230:1350–1354

Shen C-H (2019) Chapter 9—amplification of nucleic acids. In: Shen C-H (ed) Diagnostic molecular biology. Academic Press, Amsterdam, pp 215–247. https://doi.org/10.1016/B978-0-12-802823-0.00009-2

Wacker MJ, Godard MP (2005) Analysis of one-step and two-step real-time RT-PCR using SuperScript III. J Biomol Tech 16:266–271

Wittwer CT, Herrmann MG, Gundry CN, Elenitoba-Johnson KS (2001) Real-time multiplex PCR assays. Methods 25:430–442

Zumla A et al (2014) Rapid point of care diagnostic tests for viral and bacterial respiratory tract infections—needs, advances, and future prospects. Lancet Infect Dis 14:1123–1135

Western Blotting

5

Western blot (also known as protein immunoblot) is an important molecular biology technique routinely used in most research laboratories for the separation, immunodetection, and identification of proteins (Mahmood and Yang 2012; Manoussopoulos et al. 2000). Western blot was developed in 1979 by modifying available blotting techniques such as Southern blot (DNA blotting) and Northern blot (RNA blotting) (Towbin et al. 1979). In virology, western blot detects and relatively quantifies the expression of viral proteins and viral load from an infected sample.

5.1 Types of Protein Separation Techniques

There are different ways to separate a protein of interest from a pool of cellular proteins. These include:

5.1.1 Native or Nondenaturing Gel

Native or nondenaturing gel: Separates proteins without disrupting their native structure. Here, proteins are separated based on size and isoelectric point, i.e., 2-dimensional (2D) separation.

5.1.2 Denaturing Gel

Denaturing gel: Separates proteins after disrupting their native structure. Denatured proteins are separated based on size (molecular weight), i.e., 1D separation.

In either case, the separated protein can be transferred onto a solid phase and immunoblotted with an antibody specific to the protein of interest.

This chapter will focus on the most commonly used western blotting, the denaturing gel 1D analysis.

Z. Hassan, G. Khan, *Molecular Techniques for Studying Viruses*,
https://doi.org/10.1007/978-981-99-8097-0_5

5.2 Materials

PAGE, APS, TEMED, SDS, gel apparatus, rocker, imager, membrane blot, filter papers, antibodies, and protein samples.

5.3 Procedure

5.3.1 SDS-Polyacrylamide Gel Electrophoresis

The SDS-polyacrylamide gel electrophoresis (PAGE) is made up of two layers. A separating gel layer is at the bottom phase of the gel and a stacking layer at the top. To perform an SDS-PAGE, prepare a separating gel of a certain percentage of PAGE. The gel percentage depends on the molecular weight of the protein, i.e., low compact gels are required for high molecular weight proteins and vice versa.

- Dissolve appropriate PAGE (w/v) in water (or tris-SDS buffer).
- Add 10% SDS to create a negative environment that neutralises the protein net charge, so separation is based only on the size (1D).
- Add tetramethylethylenediamine (TEMED) and ammonium persulfate (APS) to stabilise free radicals and catalyse acrylamide polymerisation.
- Pour the gel into a preassembled western blot gel cassette (Fig. 5.1).
- Add approximately 1 mL of alcohol to remove bubbles, if any.
- Let the gel solidify at room temperature.
- Once solidified, discard the alcohol from the separating gel.
- Prepare a porous gel of about 4% (stacking) and pour it on the separating gel.
- Insert a gel comb of desired thickness and let it solidify.
- After solidification, remove the comb, assemble the cassette in its running holder, and place the gel running holder in a tank containing Tris-SDS buffer for electrophoresis.
- Fill the cassette running holder with the same Tris-SDS buffer and avoid leaks.
- Make a 1x mixture of protein loading dye in the purified proteins (refer to Chap. 2 for protein purification) and denature the mixture by boiling at 100 °C for 5–10 min.
- Load the denatured protein mixture into the gel wells, along with a protein marker of known molecular weight.
- Run the electrophoresis for 1–2 h at a voltage of 90–120 V (Fig. 5.1).

5.3.2 Visualisation of PAGE and Immunoblotting

Suppose that the SDS-PAGE is on a purified single protein or variants of the same protein with known molecular weight. Then the gel can be stained with dyes such as Coomassie blue or silver stain and visualised under a gel doc machine using white light. However, if the sample is a pool of proteins from total cellular proteins, for example, or if the immunoreactivity of the purified protein is needed, then immunoblotting with a specific antibody against the protein of interest has to be performed.

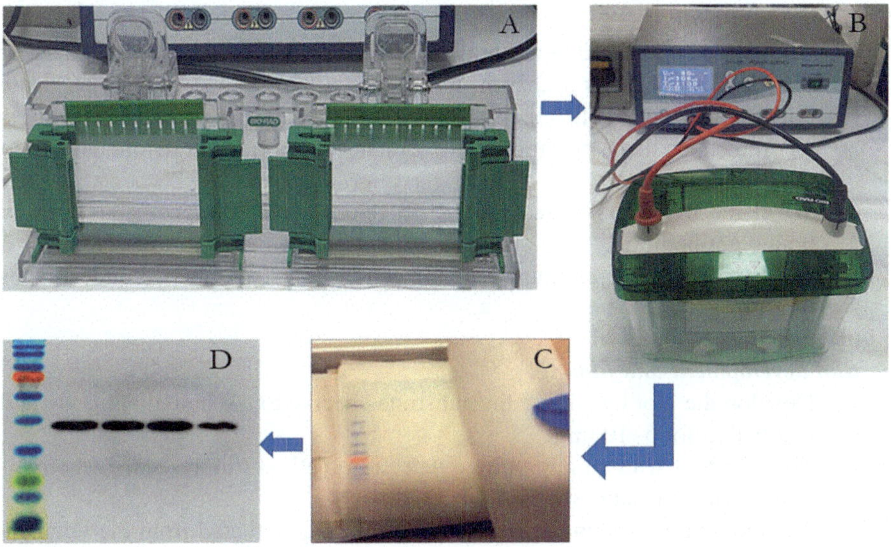

Fig. 5.1 Schematic representation of an SDS-PAGE. (**a**) SDS-PAGE gel solidifying in an assembled gel cassette stand. (**b**) Running the SDS-PAGE gel. (**c**) Transferring protein bands from gel to blot. (**d**) Western blot result

For immunoblotting, the separated proteins must first be transferred onto a membrane (nitrocellulose or PVDF) (Fig. 5.1).

1. ***Transfer protein from the gel to the membrane.***
   (a) Soak the membrane in methanol for a minute and then in the transfer buffer (Tris buffer +20% methanol) for 10–15 min.
   (b) Soak filter papers and sponge in transfer buffer for 10–15 min.
   (c) Setup an electrotransfer cassette sandwich as follows: starting with the negative side of the cassette, put a sponge→ filter papers (3 layers, for example)→ gel containing separated proteins→ a membrane blot→ 3 layers of filter papers→ sponge, and close the cassette. (**Note**: It is crucial to make the sandwich tight to avoid bubbles during transfer.)
   (d) Insert the cassette into a gel tank containing transfer buffer and apply electric current (high voltage, such as 80–90 V for a shorter time, 2–3 h, or low voltage, 30–40 V for 10–18 h). It is best to transfer at 4 °C.
   (e) After protein transfer, wash the membrane blot in saline buffer (PBS or TBS with 0.1–1% Tween-20) for 5 min with high-speed rocking.
   (f) Block any space on the membrane blot that was not occupied by proteins. Blocking is done to prevent any nonspecific binding of antibodies during immunoblotting. Blocking can be done in 3–5% BSA, nonfat dry milk (blotto) or serum, etc. (Heda et al. 2020), for at least 1 h at room temperature on a low-speed rocking.
   (g) Wash the membrane blot twice for 5 min each with high-speed rocking and continue with immunoreaction.

2. ***Immunoreaction.***
   (a) Add antibody specific to the protein of interest into the blot and incubate at room temperature for 1–2 h or at 4 °C for a longer time (10–18 h). If the optimal dilution at which the antibody should be used is not known, then it is recommended that a preliminary titration test be carried out before using it in the experiment.
   (b) Wash the blot (high-speed rocker) to remove any nonspecifically/weakly bound antibodies.
   (c) One-step or two-step immunoreaction can be done (see Chap. 5 under 'types of ELISA').
   (d) Develop the blot.
3. ***Blot development and filming.***
   (a) Develop the blot by adding the substrate specific to the bound antibody and incubating for 5–10 min.
   (b) Film the blot with an appropriate detection method (fluorescence, chemiluminescence, or radiographs).
   The amount of protein expressed can be semi-quantitated from the intensity of bands obtained from the western blotting. A relative expression can be calculated using image analysis software, such as ImageJ. The same blot can be reused for the detection of another protein that is not overlapping in size with the first one. Striping and/or re-blocking is advisable before immunoreaction for the second protein. (**Note**: A clean blot and specific monoclonal antibody are essential considerations for reblotting).
   Troubleshooting tips are highlighted (Table 5.1).
4. ***Striping of western blot membranes.***
   Stripping western blot membranes allows the detection and measurement of multiple proteins in a protein mixture that was separated in an experiment. Stripping refers to the removal of antibodies from a western blot membrane. After striping, the same blot can be re-probed for a second protein of interest using another antibody specific to that protein. **Note**: Membrane stripping may lead to the loss of some protein. Therefore, it is not appropriate to quantitatively compare proteins detected before and after stripping. For stripping western blot membrane
   (a) Incubate the membrane in stripping buffer (glycine (mild) or guanidine-tris (harsh) buffers containing 10% SDS, Tween-20, and β-mercaptoethanol) (Yeung and Stanley 2009) at room temperature for 10–60 min with gentle rocking. (**Note**: Buffer can be preheated at 50–70 °C for harsh stripping).
   (b) Discard buffer.
   (c) Repeat incubation for 5–10 min with fresh stripping buffer and gentle rocking.
   (d) Discard buffer.
   (e) Wash the membrane in a saline buffer for 10 min with high-speed rocking.
   (f) Repeat the wash step a couple of times to get rid of the stripping reagent.

Stripped membrane blot is now ready for blocking and antibody probing. Troubleshooting tips are highlighted (Table 5.1).

5.4 Troubleshooting

Table 5.1 Troubleshooting for western blotting

S/N	Problem	Possible reason(s)	Solution(s)
1.	No protein bands	• Protein has degraded, or a low quantity was loaded • The antibody used is not specific to the target protein • The primary and secondary antibodies are not compatible • Stripping buffers/ conditions are too harsh	• Perform protein assay to ensure good quality and quantity of protein before loading • Check specificity of antibodies • Ensure compatibility of primary and secondary antibodies • Do mild stripping • Optimise stripping conditions
2.	Weak bands	• Not enough protein was loaded • Inefficient antibody binding	• Increase the amount of protein loaded into the gel • Optimise antibody incubation time
3.	Nonspecific bands	• Cross-reactivity of antibodies • Nonspecific antibody binding • Protein degradation	• Check specificity of antibodies • Optimise antibody concentration and incubation time • Increase blocking time • Increase time/salt concentration or rocking speed during stringency wash • Add protease and phosphatase inhibitors to avoid partial degradation of the protein
4.	Proteins bands from the first blot are still prominent after stripping	• Stripping buffer or conditions are not appropriate	• Optimise stripping conditions based on film exposure time • Do harsh stripping • Incubate membrane at high temperature (40–50 °C)
5.	Patchy and uneven spots on the blot	• Improper transfer of proteins from gel to blot • Uneven distribution of solutions across the blot • Impurities are present in solutions, especially the secondary antibody	• Ensure no air bubbles are trapped between the gel and the membrane during sandwich preparation for electrotransfer • Constant shaking during antibody incubations and washings is crucial to ensure a uniform distribution of solutions across the blot • Centrifuge or filter secondary antibody before use

References

Heda GD, Shrestha L, Thapa S, Ghimire S, Raut D (2020) Optimization of western blotting for the detection of proteins of different molecular weight. BioTechniques 68:318–324

Mahmood T, Yang P-C (2012) Western blot: technique, theory, and trouble shooting. N Am J Med Sci 4:429–434

Manoussopoulos IN, Maiss E, Tsagris M (2000) Native Electrophoresis and Western Blot Analysis (NEWeB): a method for characterization of different forms of potyvirus particles and similar nucleoprotein complexes in extracts of infected plant tissues. J Gen Virol 81:2295–2298

Towbin H, Staehelin T, Gordon J (1979) Electrophoretic transfer of proteins from polyacrylamide gels to nitrocellulose sheets: procedure and some applications. PNAS 76:4350–4354

Yeung Y-G, Stanley ER (2009) A solution for stripping antibodies from PVDF immunoblots for multiple reprobing. Anal Biochem 389:89–91

6 Serological Assays

Serological techniques are often used for studying viruses in a liquid sample. These techniques include complement fixation, immunofluorescence, and enzyme-linked immunosorbent assay (ELISA) (Cann 2012). While they are all based on antigen–antibody (immune) reactions, they differ in their detection methods. For example, complement fixation measures the haemolysis of red blood cells in the presence of antigen–antibody complexes. In contrast, immune complexes in immunofluorescence and ELISA are detected through fluorescent emission and enzymatic reaction, respectively (Fig. 6.1). This section will describe the use of ELISA as it is one of the most popular techniques used to detect viruses in liquid samples such as blood.

ELISA is an immunoassay that employs an enzymatic reaction for the detection of immune complexes. Results of ELISA are measured from the signal produced by the substrate specific to the enzyme conjugated to the detection antibody (Pandey et al. 2019). Based on the optical densities produced, ELISA can also estimate the quantities of viral antigen(s) or antibodies. There are several types of ELISAs (Fig. 6.1).

6.1 Types of ELISA

6.1.1 Direct ELISA

Direct ELISA involves the immobilisation of a virus-infected sample onto a plate and detection using an enzyme-linked antibody directed against a viral antigen. Direct ELISA utilises a single antibody to detect both antigen and signal. Direct ELISA is also referred to as a one-step approach.

Z. Hassan, G. Khan, *Molecular Techniques for Studying Viruses*,
https://doi.org/10.1007/978-981-99-8097-0_6

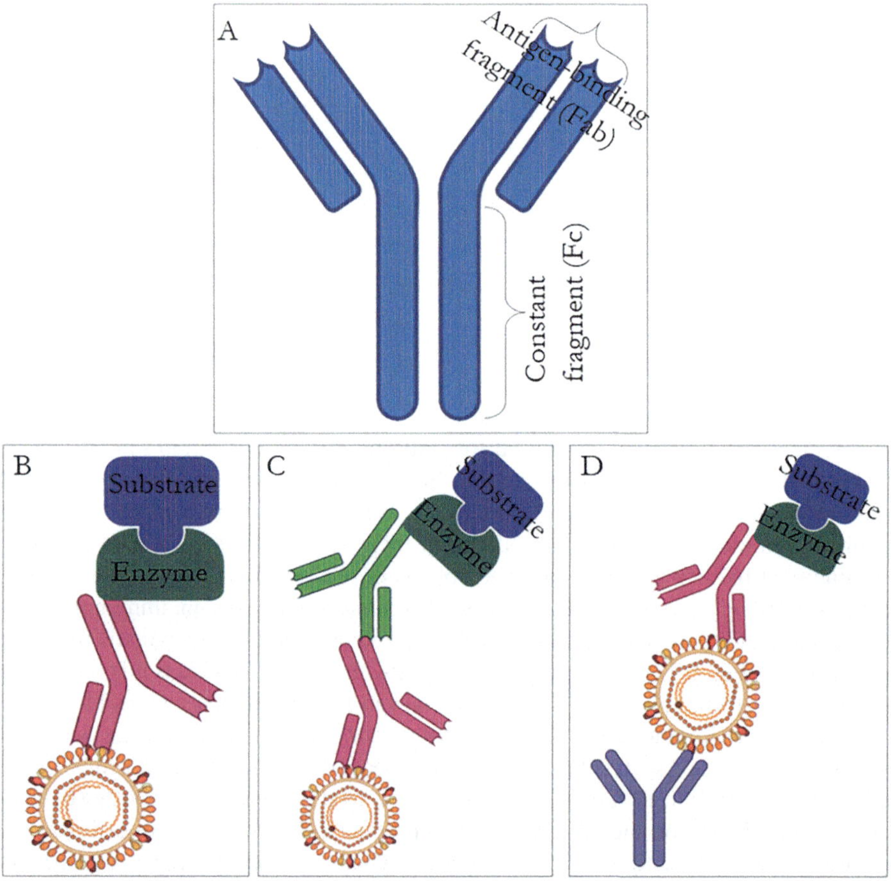

Fig. 6.1 Types of ELISA. (**a**) Structure of an antibody showing the Fab and Fc regions. (**b**) Direct ELISA. (**c**) Indirect ELISA. (**d**) Sandwich ELISA. (Image created in BioRender)

6.1.2 Indirect ELISA

Indirect ELISA involves detecting viral antigen from an immobilised sample using an antigen-specific primary antibody followed by an enzyme-linked secondary antibody specific to the Fc portion of the primary antibody. Here, two antibodies are used: one for antigen detection and another for signal detection. Indirect ELISA is also referred to as a two-step approach.

6.1.3 Sandwich ELISA

Sandwich ELISA involves immobilising a viral antigen-specific (capture) antibody onto a 96-well plate before adding the virus-infected sample into the wells. Then,

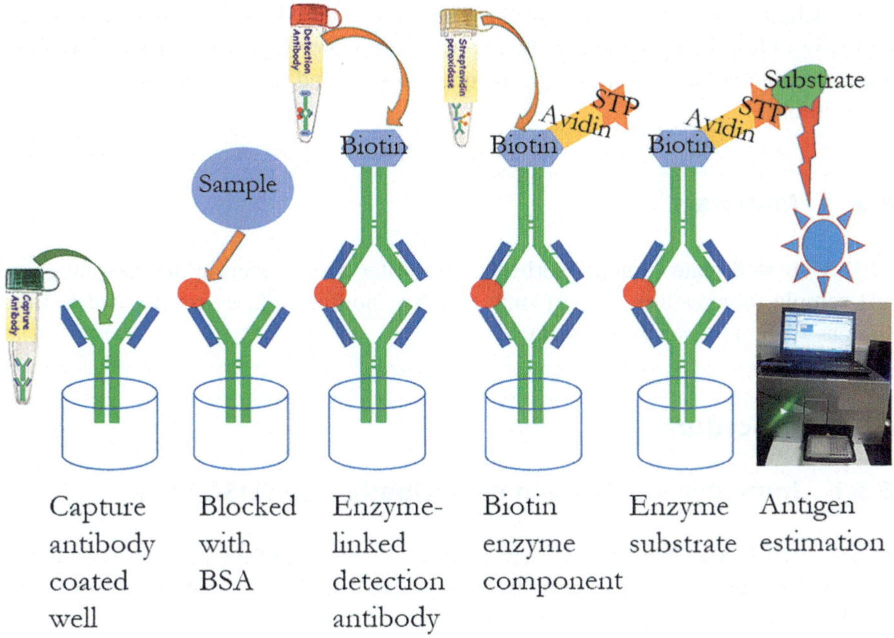

Fig. 6.2 Schematic representation of a sandwich ELISA. Sandwich ELISA with one-step (direct) detection. (Image adapted from (Pandey et al. 2019) with modification)

another viral antigen-specific antibody targeting a different epitope is added to the reaction. Hence, the viral antigen is sandwiched between two antibodies (Fig. 6.2). Sandwich ELISA can be adapted to a one-step or two-step approach.

6.1.4 Competitive ELISA

Competitive ELISA involves immobilising a capture antibody onto a plate and then adding a virus-infected sample and a protein competitor at the same time. A protein competitor is an antigen that is specific to the capture antibody and conjugated to a substrate. (**Note**: The capture antibody preferentially binds to the viral antigen with higher affinity than the protein competitor.) The signal for competitive ELISA is developed by targeting the substrate conjugated to the protein competitor. Thus, unlike in the other types of ELISA described above, the signal generated in competitive ELISA is indirectly proportional to the amount of viral antigen in the sample. Competitive ELISA can also be adapted to a one-step or two-step approach.

A reverse protocol can be used for competitive ELISA, i.e., immobilising an unconjugated protein competitor onto the wells, then adding a virus-infected sample and a labelled antibody to the wells. The sample can be pre-incubated with the antibody before being added to the capture competitor (*optional*). Therefore, only the unbound antibody from the pre-incubation can bind to the immobilised antigen competitor.

6.2 Materials

ELISA 96-well plate (immunosorbent), microtiter plate reader, antibodies/antigens, test sample, nonreacting protein such as BSA, nonfat milk, or casein, and buffers such as PBS, TBS, or bicarbonate.

6.3 Procedure

6.3.1 Immobilising Antigen or Antibody into ELISA Plate

- To coat ELISA 96-well plate, add a buffered solution of the antigen or antibody into the wells. (A solution of 0.05 M sodium carbonate–sodium bicarbonate, pH 9.6, is used as ELISA coating buffer.)
- Incubate at 4 °C overnight or 37 °C for 2–3 h for the antigen or antibody to adhere to the plastic through charge interactions (Bantroch et al. 1994).
- Wash off the unbound antigen/antibody in saline solution (PBS Tween or TBS-Tween).
- Add a nonreacting protein (BSA, for example) to block any unoccupied surface in the wells.
- Incubate the plate for 10–15 min or up to 1 h.
- Wash off the solution.
- Coated plate is ready for use. It can be stored at 4 °C or − 20 °C for later use.

Other immobilisation methods such as dehydration, chloroform-ethanol, etc., can also be used (Bantroch et al. 1994). **NOTE**: Viral antigens used for coating wells of a 96-well plate can be isolated from infected cells or tissues either as a crude or purified single protein (see Chap. 2 under 'protein isolation'). A microbial culture, for example, *E. coli,* can be used for mass production of viral proteins.

6.3.2 Conjugating an Antibody or Antigen

- Mix adequate concentration (e.g., 1–10 mM) of the antibody (or antigen) and the tag molecule such as biotin, horseradish peroxidase, or streptavidin peroxidase.
- Incubate the mixture at room temperature for 2–3 h with shaking.

- Remove free tag proteins by dialysis, gel-filtration, or ultrafiltration (Wiener et al. 2020).
- The concentration of the resulting conjugated antibodies (or antigen) is determined by the absorbance at 260 nm using spectroscopy (Wiener et al. 2020) (see Chap. 2 under 'determination of protein purity and concentration').
- Conjugates can be stored at 4 °C or − 20 °C for later use.

6.3.3 Detection of Viral Antigen or Antibody

- Add a sample containing viral antigen/antibody into the respective pre-coated wells and incubate at room temperature or 37 °C for 1–2 h for the immune reaction to occur.
- Wash the wells to remove any unbound antibody/antigen.
- Add antibody based on the choice of ELISA type (see above under 'types of ELISA').
- Wash twice to remove the unbound antibody.

6.3.4 Signal Detection

- Add appropriate substrate for the enzyme.
- Incubate the reaction for 10–30 min at room temperature for the signal to develop.
- Stop the reaction by adding the stop solution specific to the substrate used.
- Read and measure the results using a microtiter plate reader.

Note: The result of ELISA depends on the intensity of the substrate. Therefore, monitor colour change if the signal is chromogenic to avoid overdevelopment. Chromogenic substrates can be visualised and semi-quantified by the eyes. Quantitative measurement is achieved by a microtiter plate reader. There are fluorescent and electrochemical signals that can be used with ELISA (Pandey et al. 2019).

Troubleshooting tips are highlighted (Table 6.1).

6.4 Troubleshooting

Table 6.1 Troubleshooting for ELISA

S/N	Problem	Possible reason(s)	Solution(s)
1.	Signal in the blank well (high background)	• Improper blocking • Presence of unbound materials in the well	• Increase blocking time or concentration of blocking solution • Increase the number of washings after each step
2.	No or poor signal	• Improper antibody compatibility • Inadequate reaction/ development • Inappropriate plate	• Ensure compatibility of primary and secondary antibodies in two-step ELISA • Optimise antibody concentration • Increase incubation time • Increase development time • Ensure the use of ELISA-friendly plates
3.	Too much signal (false positive)	• Overdevelopment	• Decrease development time • Optimise antibody and substrate concentration
4.	Poor assay-to-assay reproducibility	• Inconsistency in protocol between assays	• Check overall protocol between the assays • Increase washings • Use fresh buffers • Check negative controls

References

Bantroch S, Bühler T, Lam JS (1994) Appropriate coating methods and other conditions for enzyme-linked immunosorbent assay of smooth, rough, and neutral lipopolysaccharides of Pseudomonas aeruginosa. Clin Diagn Lab Immunol 1:55–62

Cann A (2012) Principles of molecular virology. Academic Press, Amsterdam

Pandey AK, Varshney RK, Sudini HK, Pandey MK (2019) An improved Enzyme-Linked Immunosorbent Assay (ELISA) based protocol using seeds for detection of five major peanut allergens Ara h 1, Ara h 2, Ara h 3, Ara h 6, and Ara h 8. Front Nutr 6:68

Wiener J, Kokotek D, Rosowski S, Lickert H, Meier M (2020) Preparation of single- and double-oligonucleotide antibody conjugates and their application for protein analytics. Sci Rep 10:1457

Immunoprecipitation 7

Immunoprecipitation (IP) is an immunological technique in which antibodies are used to pull down a target protein from a mixed solution. This happens through affinity binding between the antibody and protein of interest. IP technique helps to identify in vivo protein–protein and protein–nucleic acid interactions (Ranawakage et al. 2019). Once the protein of interest is precipitated from the solution, the interacting protein(s) or nucleic acid (DNA or RNA) can be identified by western blot or PCR, respectively. There are two main categories of IP. These are:

1. **Native IP**: That is when the cell lysate is untreated. Native IP is mainly for the immunoprecipitation of nucleic acid-binding proteins that are either highly expressed or have a strong affinity towards the target DNA/RNA molecules.
2. **Crosslinked IP**: When the cells are treated with crosslinkers to stabilise the nucleic acid–protein interactions before lysis. Crosslinking is done to prevent the pull down of nucleic acids that associate with the protein after cell lysis (Mukherjee et al. 2021). This type of IP helps to identify molecules that have transient or weak affinity to the protein of interest, even after stringent laboratory handling. Example of crosslinking agents frequently used in IP include UV radiation, formaldehyde, etc. Formaldehyde can also stabilise protein–protein interactions in addition to crosslinking nucleic acid to protein molecules. Therefore, the use of formaldehyde can increase the detection of indirect protein interactions.

7.1 Types of Immunoprecipitation

7.1.1 Individual Protein Immunoprecipitation

Individual protein immunoprecipitation is the use of immunoreaction to pull down a specific protein of interest from a crude lysate solution. It is often performed with liquid samples such as plant tissues, body fluids including blood, cerebrospinal fluid, urine etc. This technique is a good alternative to ELISA.

Z. Hassan, G. Khan, *Molecular Techniques for Studying Viruses*,
https://doi.org/10.1007/978-981-99-8097-0_7

7.1.2 Chromatin Immunoprecipitation

Chromatin immunoprecipitation (ChIP) assay is used to precipitate DNA–protein interactions. It identifies the regions of the genome that are associated with a particular protein or the proteins that associate with a specific region of the genome. This technique also reveals proteins or protein modification(s) that occur at a particular region of the genome at a given time or stage of development. Figure 7.1 shows the graphical flow of ChIP protocol.

7.1.3 RNA Immunoprecipitation

RNA immunoprecipitation (RIP) assay simply refers to immunoprecipitation that aims to identify RNA-binding protein (RBP). These RNAs can be mRNAs or noncoding RNAs.

7.1.4 Co-Immunoprecipitation

Co-immunoprecipitation (Co-IP) is the pull-down of a protein along with any molecule or ligand that is bound to it. Co-IP often targets a protein believed to be a member of a complex protein family, such as Agrobacterium type VI secretion system (Lin and Lai 2017). Here, an antibody specific to the protein of interest is used to capture and precipitate the protein of interest along with its binding proteins using a resin (Tang and Takahashi 2018).

7.2 Materials

Tissue culture flask, pipettes, incubator, heat block, centrifuge, tubes, magnetic separator stand and rack, crosslinker, vortex machine, end-to-end rotator, cells, media, magnetic beads, antibodies, RNA isolation and reverse transcription reagents, protein isolation reagents, PCR reagents, western blot reagents.

7.3 Procedure

7.3.1 Lysate Preparation

- Harvest viral infected cells by centrifugation.
- Resuspend cell pellet in cold crosslinking solution (10% formalin, 4% paraformaldehyde) and incubate for 2–4 h at room temperature.
- Wash the crosslinked cells twice in 1 × PBS for 5 min each.
- Resuspend the cells in 100 mL lysis buffer (50 mM HEPES, 140 mM NaCl, 1 mM EDTA, 1% Triton X, and 0.1% sodium deoxycholate) supplemented with protease inhibitor, and incubate at 4 °C for 30 min, then store at −80 °C or liquid nitrogen.

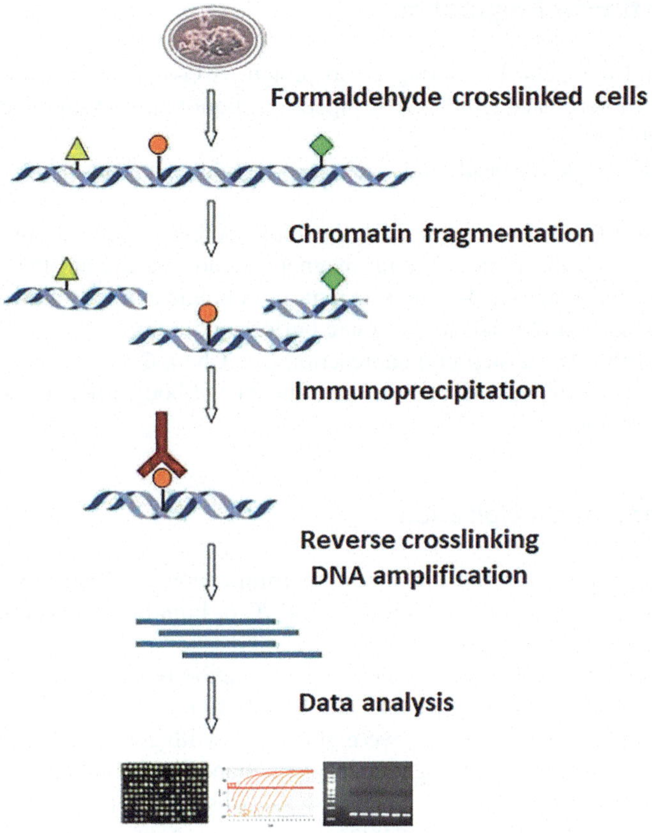

Fig. 7.1 ChIP graphical workflow

Alternatively, physical methods of cell lysis such as sonication, mechanical disruption by grinding with mortar and pestle or enzymatic methods can be used.

- Thaw the lysate on ice. Repeat freeze-thaw cycles 2–3 times.
- Store the lysate at −80 °C or liquid nitrogen until use.
- *Optional*: Fragment the genome into smaller units for ChIP. Fragmentation may not be required for RIP.
- Centrifuge the cell lysate at high speed (~13,000 × g) 4 °C for 15–20 min to pellet cellular debris.
- Transfer the supernatant to a new tube and place it on ice and continue with immunoprecipitation.
- *Optional*: Perform Bradford assay to determine the protein concentration of the lysate.

7.3.2 Antibody Preparation

- Resuspend magnetic beads (tagged to protein A/G-agarose beads or indirectly bound to beads coated with IgG) by pipetting, aliquot into 1.5 mL tubes, and wash couple of times.
- Add antibodies to the beads and incubate overnight on an end-to-end rotator at 4 °C.
- Add 1 mL of lysis buffer and wash the antibody-bead conjugate by rotating for 5 min at 4 °C. Place the tubes on magnetic stand and discard the supernatant. Repeat the washing step 4–5 times to remove unbound antibody and to equilibrate the antibody-bound beads to the lysate buffer conditions.
  Alternatively, do a 2-step immunoreaction (see Chap. 6).
- *Optional*: Perform western blot to determine the antibody efficiency and optimise incubation time.

7.3.3 Immunoprecipitation

- Take 10% of the cell lysate into a new microfuge tube and label it as 'input'. Add TRIzol reagent, mix well and freeze at −80 °C or liquid nitrogen until use during nucleic acid isolation step.
- To the remaining 90% cell lysate, add the antibody of choice conjugated to magnetic beads and top it to 1 mL with lysis buffer.
- Incubate the mixture for 1 h or overnight at 4 °C with gentle end-to-end rotation.
- Purify the sample by separating it on magnetic stand and aspirating the supernatant.
- Wash the immunoprecipitated beads with salt wash buffer (e.g., 1 × PBS) several times (5–8 times) by gently vortexing and separating on a magnetic stand.
- Decrosslink by heating the mixture at 95 °C for 15–20 min.
- Purify DNA/RNA by resuspending the sample in 100 μL of 1 × PBS supplemented with 0.5 μL proteinase K and 1 μL of RNasin Plus RNase inhibitor.
- Incubate at 42–55 °C for 30 min with end–end rotation.
- Spin down the samples and place them on a magnetic rack for 2 min.
- Collect the supernatant into a new tube as immunoprecipitated nucleic acid and add TRIzol reagent.
- Isolate nucleic acid from both input and immunoprecipitated samples, as described in Chap. 2.
- Identify the nucleic acid by PCR or sequencing.

Troubleshooting tips are highlighted (Table 7.1).

7.4 Troubleshooting

Table 7.1 Troubleshooting for immunoprecipitation

S/N	Problem	Possible reason(s)	Solution(s)
1.	No band of the protein of interest on western blot after cell lysis	• Antibody is not suitable for the protein of interest • Poor quality or specificity of antibodies • The primary and secondary antibodies are not compatible	• Check the specificity of antibodies using western blot • Optimise antibody concentration and incubation time • Check the specificity of antibodies • Ensure compatibility of primary and secondary antibodies
2.	Weak western blot bands	• Not enough protein was loaded • Inefficient antibody binding	• Increase the amount of protein loaded into the gel • Optimise antibody incubation time
3.	Nonspecific western blot bands	• Cross-reactivity of antibodies • Nonspecific antibody binding. • Indirect interactions • Protein degradation	• Check the specificity of antibodies • Subject the immunoprecipitated products to stringent washings • Optimise antibody concentration and incubation time • Check and optimise crosslinker • Use fresh cell lysate • Optimise cell lysis and fragmentation
4.	Low signal-to-noise ratio	• Too much lysate during IP • Nonspecific RNA binding to the beads	• Decrease the amount of lysate used • Coat beads with secondary antibody and perform 2-step immunoreaction • Do a sandwich antibody-bead conjugation (two specific antibodies to the protein of interest)
5.	Weak or no PCR amplification	• Transient or weak interaction between the nucleic acid and the protein of interest	• Crosslink cells to stabilise the protein–nucleic acid interactions before lysis
6.	Low nucleic acid yield	• DNase/RNase contamination	• Ensure all instruments, workspaces, and reagents are nuclease-free

References

Lin J-S, Lai E-M (2017) Protein-protein interactions: co-immunoprecipitation. Methods Mol Biol 1615:211–219

Mukherjee P, Kurup RR, Hundley HA (2021) RNA immunoprecipitation to identify in vivo targets of RNA editing and modifying enzymes. Methods Enzymol 658:137–160

Ranawakage DC, Takada T, Kamachi Y (2019) HiBiT-qIP, HiBiT-based quantitative immunoprecipitation, facilitates the determination of antibody affinity under immunoprecipitation conditions. Sci Rep 9:6895

Tang Z, Takahashi Y (2018) Analysis of protein–protein interaction by Co-IP in human cells. In: Methods in molecular biology, vol 1794. Humana Press, Clifton, N.J, pp 289–296

8 Small Interfering RNA

The central dogma of molecular biology states that DNA is transcribed to RNA, which in turn is translated to protein. However, only about 2% of RNAs code for proteins. About 98% of the human transcripts are estimated to be non-protein coding, hence referred to as noncoding RNAs. Among the noncoding RNAs are the small noncoding RNAs, known as microRNAs (miRNAs). These RNAs are around 18–25 nucleotides in length and are usually single-stranded. However, they can be double-stranded and have hairpin structures (Zhao et al. 2015). microRNAs are generated from endogenous transcripts (Zhao et al. 2015; Fire et al. 1998). The biogenesis miRNA is shown in Fig. 8.1. miRNAs have the ability to interfere with gene expression at both transcriptional and post-transcriptional (via translation inhibition) levels (Hamilton and Baulcombe 1999; Bernstein et al. 2001) (Fig. 8.1). Thus, they are silencing gene expression via the RNA interference (RNAi) pathway.

RNA interference also known as post-transcriptional gene silencing is a naturally occurring conserved biological response to double-stranded RNA that mediates gene expression (RNA Interference (RNAi) n.d.). It was discovered by Fire and colleagues in 1998 (Fire et al. 1998). RNAi acts on both endogenous and exogenous nucleic acids, including viral RNAs. The mechanism for RNAi is sequence complementarity. The discovery of this pathway has revolutionised experimental biology in many aspects of biological sciences, including functional genomics, understanding viral infection and pathogenesis, therapeutic intervention, and lot more.

The RNAi pathway is widely utilised in the laboratory through in vitro synthesis of small RNAs termed as small interfering RNA (siRNA). siRNAs are mostly of exogenous origin, i.e., in vitro synthesised or processed short RNAs that mimic the naturally occurring miRNAs (Elbashir et al. 2001a, b). However, they can be produced naturally (Bernstein et al. 2001). Unlike the miRNAs that have broad spectrum of action, as signified by their ability to accommodate few mismatches, siRNA requires 100% sequence complementarity. siRNA gene silencing is often

Z. Hassan, G. Khan, *Molecular Techniques for Studying Viruses*,
https://doi.org/10.1007/978-981-99-8097-0_8

transient since they are not self-replicating. In this chapter, we will discuss the biogenesis and mechanism of action of miRNA as it also applies to siRNA.

8.1 Biogenesis and Mechanism of Action of miRNA

- Primary transcripts of miRNA (pri-miRNA) are transcribed by RNA polymerase II and processed by RNase III (Zhao et al. 2015).
- These pri-miRNAs are bound and processed by an enzyme complex called DROSHA, forming precursor miRNAs (pre-miRNA) that are then transported from the nucleus to the cytoplasm for maturation. **Note**: siRNAs are often transfected as dsRNA with short loop between the two strands. This serves as pre-miRNA molecule.
- These pre-miRNAs bind to an enzyme complex called Dicer, which cleaves the dsRNA into 21–25 nt dsRNA.
- AGO_2 protein links the matured miRNA to the RNA-induced silencing complex (RISC) where one of the miRNA strands—the sense strand (also known as the passenger strand) is eliminated and the antisense strand serves as a guide to link the RISC to a target mRNA.
- The guide miRNA binds to its target mRNA by sequence complementarity and induces mRNA cleavage. Thereafter, the resulting cut mRNA is recognised as

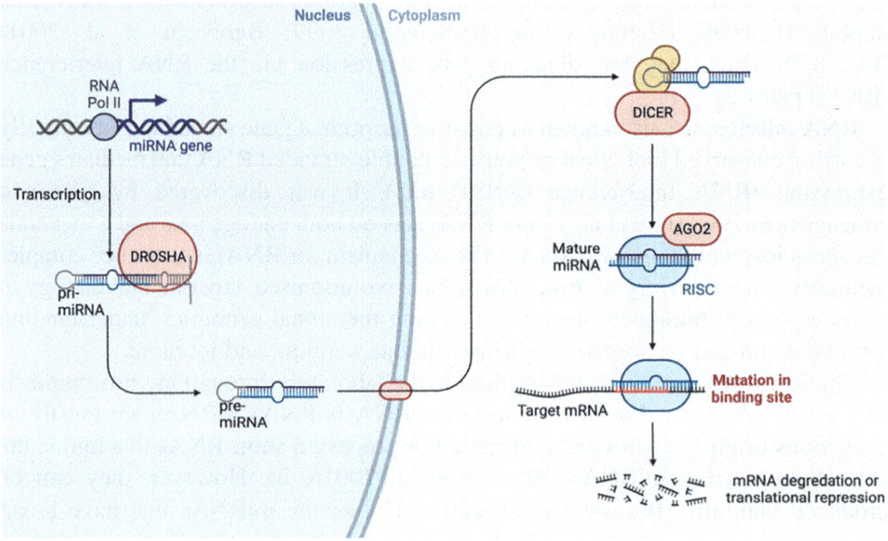

Fig. 8.1 Biogenesis of miRNA. (Image adapted from Biorender templates with slight modifications)

abnormal and then degraded by cellular exonucleases. And so, represses the expression of gene encoded by the mRNA.

8.2 Types of siRNA Transfection Techniques

Since siRNAs are often introduced into cells by transfection, we will discuss the three most common transfection methods used. The main groups of siRNA nanovectors include viral vectors, lipid-based, non-lipid organic-based, and inorganic vectors.

8.2.1 Viral-Based Transfection Technique

Understanding viral infection leads to the manipulation of the virus to deliver siRNA in cells. Viruses used for siRNA delivery include adenovirus, adeno-associated virus retrovirus, alphaviruses, flaviviruses, and Sendai virus (SeV) (Lundstrom 2020; Wang et al. 2017). These viral vectors are non-replicating. Others are conditionally replicating lentivirus self-replicating rhabdoviruses and alphaviruses (Lundstrom 2020). Some of these vectors, such as self-replicating RNA viruses, e.g., influenza A virus and Borna disease viruses are targets for therapeutic interventions since they could induce long-term transcriptional gene silencing (Lundstrom 2020; Baltusnikas et al. 2009).

8.2.2 Lipid-Based siRNA Transfection Technique

Lipid-based nanovectors are excellent for delivering siRNA into circulation and solid tumours due to their excellent biocompatibility, low toxicity, and low immunogenicity (Yonezawa et al. 2020). Other advantages of using lipid-based vectors for siRNA delivery include structural flexibility and ease of large-scale preparation (Yonezawa et al. 2020). Additionally, they are easily degradable in the blood and have negative-charge density (Yonezawa et al. 2020). Lipid-based vectors are used as excellent carriers of siRNA and other nucleic acids for delivery into complex biological systems and also in therapeutics (Yonezawa et al. 2020; Zhang et al. 2012).

Practically, lipid-based nanoparticles are prepared by dissolving a lipid mixture in an organic solvent such as chloroform and methanol and then incubate at 37 °C for 10–30 min. Then, add siRNA to the lipid mixture. Remove the organic solvent from the mixture by evaporation or dialysis in water. Now, siRNA-loaded lipid nanovectors are ready for transfection. Examples of lipid-based nanovectors used in the delivery of siRNA include liposomes, Lipidoid nanoparticles, stable nucleic acid-lipid particles (SNALPs), multifunctional envelope-type nano device (MEND) system, stable nucleic acid-lipid particles, SS-cleavable and pH-activated lipid-like material (ssPalm), and exosomes.

8.2.3 Non-lipid Organic-Based siRNA Technique

There are other nanovectors prepared in organic solvents for siRNA delivery that are not lipid-based. These non-lipid-based organic nanovectors are developed in a way that they combine the unique features of the different types of non-lipid-based vectors along with some affinity moiety conjugated to the surface of the vectors for improved uptake. These combinations were shown to improve tumour cell uptake and do not seem to alter their biodistribution (Shen et al. 2012). The non-lipid-based organic nanovectors include cyclodextrin-based nanoparticles, chitosan, polyethylenimines, and dendrimers.

8.2.4 Inorganic siRNA Technique

Gold nanoparticles are being used as delivery vectors for siRNA due to their good biocompatibility, high transfection efficiency, and lack of cytotoxicity (Guo et al. 2010; Lei et al. 2020). The gold particle captures siRNA through either electrostatic interaction or thiol linkages. The intracellular release of the siRNA content in the gold nanovectors can be triggered by glutathione, pH, or external stimuli, such as light (Guo et al. 2010). These gold nanoparticles are practically easy to synthesise and have monodispersity properties (Guo et al. 2010).

8.3 Materials

Tissue culture flask, tissue culture plate, pipettes, incubator, centrifuge, heat block, spectrophotometer, centrifugal evaporator, cells, media, gene-specific siRNA, non-specific (scrambled) RNA, antibodies, transfection reagents, RNA isolation, reverse transcription reagents, protein isolation reagents, PCR reagents, western blot reagents.

8.4 Procedure

8.4.1 Design and Preparation of siRNA

- Choose a gene of interest and design its siRNA sequences (both sense and antisense). There are some in silico siRNA design tools available online. These include http://sidirect2.rnai.jp/; https://www.genscript.com/design_center.html; https://eurofinsgenomics.eu/en/ecom/tools/sirna-design/.
- Synthesise the siRNA sequences (both sense and antisense).

- Get rid of the protecting groups used during siRNA synthesis by dissolving the siRNAs in 2′-deprotection buffer and vortex for 10 s, then incubate at 60 °C for 30 min, and then dry the siRNAs at room temperature using centrifugal evaporator (Li and Zamore 2019).
- Dissolve the siRNA in nuclease-free water and measure its concentration using spectrophotometer.
- Prepare annealing reaction by mixing equal moles of sense siRNA, and antisense siRNA (2 nmol for example) in annealing buffer (total concentration of 0.1– 1×), top up with nuclease-free water to make up 100 μL.
- Incubate the annealing reaction at 95 °C for 5 min and then for 2 h at 37 °C.
- Confirm the successful preparation of the siRNA by performing native PAGE using the siRNA duplexes (dilute 2 μM in 1× native gel-loading buffer), sense siRNA separate (dilute 4 μM in 1× native gel-loading buffer), and antisense siRNA separate (dilute 4 μM in 1× native gel-loading buffer). **Note**: siRNA duplexes should migrate slightly slower than either DNA markers of the same length or single-stranded (sense and antisense) siRNA controls.
- Now, the siRNA solution is ready. It can be used immediately or stored at −20 °C for short-term storage or at −80 °C for long-term storage.

8.4.2 Transfection of siRNA into Cells and Gene Silencing

- In a tissue culture plate, grow cells overnight in 1 mL complete media without antibiotics.
- Replace the media with 2 mL of serum-free media and incubated for 24 h.
- Prepare the siRNA reagent by diluting the siRNA specific to the target gene in the transfection media.
- After growth in serum-free media for 24 h, replace the media with the siRNA transfection media and incubate for 48 h. **Note:** Always use at least one nonsilencing (scrambled) siRNA and no siRNA (negative control) controls in the same experiment with the silencing siRNA.
- Next, harvest the siRNA-treated cells (as well as the control cells) by pipetting up and down and then centrifuge at medium speed (around 2000 × g) for 10 min.
- Discard the supernatant and collect the cell pellet for the RNA and protein isolations.
- See Chaps. 2 and 3 for RNA and protein isolation, respectively.

Troubleshooting tips are highlighted (Table 8.1).

8.5 Troubleshooting

Table 8.1 Troubleshooting for siRNA

S/N	Problem	Possible reason(s)	Solution(s)
1.	Both RNA and protein are expressed at the same rate in siRNA-treated cells and control cells	• Silencing did not happen • Scrambled RNA may be silencing the protein target	• Check the specificity of the siRNA used • Increase the concentration of siRNA • Check the specificity of the scrambled RNA used
2.	No RNA or protein is detected	• No or poor quality of RNA or protein in the sample	• Check RNA and protein isolation protocols • Ensure the process of PCR and western blot used for the detection
3.	Cells death	• Over silencing of an essential protein	• Decrease siRNA concentration • Optimise the incubation time
4.	Poor assay-to-assay reproducibility	• Common with transient transfection	• Collect 2–3 wells of technical replicates in one tube, mix well and then aliquot. Use the mean of the aliquots for analysis • Always perform experiments with test siRNA and control RNA at the same time and on the same plate • Compare the outcome of the same experiment

References

Baltusnikas J, Satkauskas S, Lundstrom K (2009) Constructing RNA viruses for long-term transcriptional gene silencing. Trends Biotechnol 37:20

Bernstein E, Caudy AA, Hammond SM, Hannon GJ (2001) Role for a bidentate ribonuclease in the initiation step of RNA interference. Nature 409:363–366

Elbashir SM et al (2001a) Duplexes of 21-nucleotide RNAs mediate RNA interference in cultured mammalian cells. Nature 411:494–498

Elbashir SM, Lendeckel W, Tuschl T (2001b) RNA interference is mediated by 21- and 22-nucleotide RNAs. Genes Dev 15:188–200

Fire A et al (1998) Potent and specific genetic interference by double-stranded RNA in Caenorhabditis elegans. Nature 391:806–811

Guo S et al (2010) Enhanced gene delivery and siRNA silencing by gold nanoparticles coated with charge-reversal polyelectrolyte. ACS Nano 4:5505–5511

Hamilton AJ, Baulcombe DC (1999) A species of small antisense RNA in posttranscriptional gene silencing in plants. Science 286:950–952

Lei W-X et al (2020) Construction of gold-siRNANPR1 nanoparticles for effective and quick silencing of NPR1 in Arabidopsis thaliana. RSC Adv 10:19300–19308

Li C, Zamore PD (2019) Preparation of siRNA duplexes. Cold Spring Harb Protoc 2019:4. https://doi.org/10.1101/pdb.prot097444

Lundstrom K (2020) Viral vectors applied for RNAi-based antiviral therapy. Viruses 12:924

RNA Interference (RNAi) (n.d.) https://www.ncbi.nlm.nih.gov/probe/docs/techrnai/

Shen H, Sun T, Ferrari M (2012) Nanovector delivery of siRNA for cancer therapy. Cancer Gene Ther 19:367–373

Wang H et al (2017) Significant inhibition of Tembusu virus envelope and NS5 gene using an adenovirus-mediated short hairpin RNA delivery system. Infect Genet Evol 54:387–396

Yonezawa S, Koide H, Asai T (2020) Recent advances in siRNA delivery mediated by lipid-based nanoparticles. Adv Drug Deliv Rev 154:64–78

Zhang S, Zhi D, Huang L (2012) Lipid-based vectors for siRNA delivery. J Drug Target 20:724–735

Zhao J et al (2015) High-throughput sequencing of RNAs isolated by cross-linking immunoprecipitation (HITS-CLIP) reveals Argonaute-associated microRNAs and targets in Schistosoma japonicum. Parasit Vectors 8:589

Histological Methods

9

Histology is the branch of science that deals with the study of tissues. Viral products, including proteins and nucleic acids (DNA or RNA), can be detected in histological samples. These can be clinical tissue samples (e.g., biopsies, post-mortem tissues) or laboratory-maintained infected cells (cell culture). Histological techniques can be combined or performed individually. For example, viral proteins and nucleic acids can be detected in the same tissue section (Renshaw 2017). In contrast to liquid samples (e.g., blood, urine, CSF, pulmonary aspirate), tissue analysis can provide additional information on cellular morphology and tissue architecture. Thus, histology can reveal the extent of cytopathic effects and cell tropism of the virus.

9.1 Types of Histological Techniques

1. **Immunohistochemistry (IHC)**: can be used to detect viral antigens in infected tissue samples. It is based on the principles of immunoassays, which involve using specific antibodies that target epitopes on the antigens. These antigens can be viral or cellular proteins (Reguraman et al. 2021). IHC is also widely used to identify and localise viral infections (Reguraman et al. 2021). IHC is an essential part of histopathology, routinely used in diagnosis (Renshaw 2017). It helps to characterise the different cellular components that constitute a tissue and identify the pathological changes that happen due to the viral infection. For example, IHC is used for diagnosing and characterising diseases. These include malignancies, such as lymphomas and autoimmune disorders like multiple sclerosis (Hassani et al. 2021).
2. **Immunocytochemistry (ICC)**: can be used to detect viral antigens in infected cells using specific antibodies. The main difference between ICC and IHC is that ICC is performed on cells or swabs where their surrounding extracellular matrix is removed. ICC is commonly used as a rapid diagnostic method for detecting

Z. Hassan, G. Khan, *Molecular Techniques for Studying Viruses*,
https://doi.org/10.1007/978-981-99-8097-0_9

viruses such as influenza or respiratory syncytial virus in nasopharyngeal swabs (Hopkins et al. 2003; Treuhaft et al. 1985).
3. **In situ hybridisation** (ISH): can be used to detect viral nucleic acids within an infected tissue or cells. ISH uses probes that are complementary to the target viral DNA/RNA sequences (Khan et al. 1992). Probes are usually labelled (end-labelled or nick-labelled) with markers such as biotin or digoxigenin. The marker is then targeted using labelled antibodies for immunoreaction.

IHC, ICC, and ISH are performed in three fundamental steps: sample preparation, immunoreaction, and signal detection. (**Note:** If the probe is tagged to a molecule conjugated directly to a detection molecule, the immune reaction step is not needed in ISH.)

9.2 Materials

Embedding station, incubator, oven, refrigerator, freezer, microtome, microscope, microwave, coated slides, coverslips, water bath, wax, xylene, ethanol, probes, antibodies, developers, counter-stain, mountants and test samples tissue/cells, saline buffers such as saline-sodium citrate (SSC).

9.3 Histological Sample Preparation

Histological samples are prepared in various ways based on the nature of the sample and its intended purpose. Here, we will highlight some methods of sample preparation commonly employed in virology laboratories.

9.3.1 Cytospin

A cytospin, also called cytocentrifuge, is a specialised centrifuge that spins down and concentrates cells onto a microscope slide for histological examination (Fig. 9.1). This instrument is essential for histological examinations of liquid specimens with low cell densities, such as clinical fluid specimens. These include blood, urine, sperm, bronchoalveolar lavage, and cell culture. **Note**: Centrifugal force can affect cells' architecture. Therefore, it is essential to perform cytospins at low speeds. After cytospin, a cell smear must be processed immediately to preserve the cellular details (Brunzel 2016).

9.3.2 Cryostat

A cryostat is an instrument equipped with a microtome and maintains cryogenic temperature (−15 °C to −30 °C) (Fig. 9.1). This temperature helps to preserve the

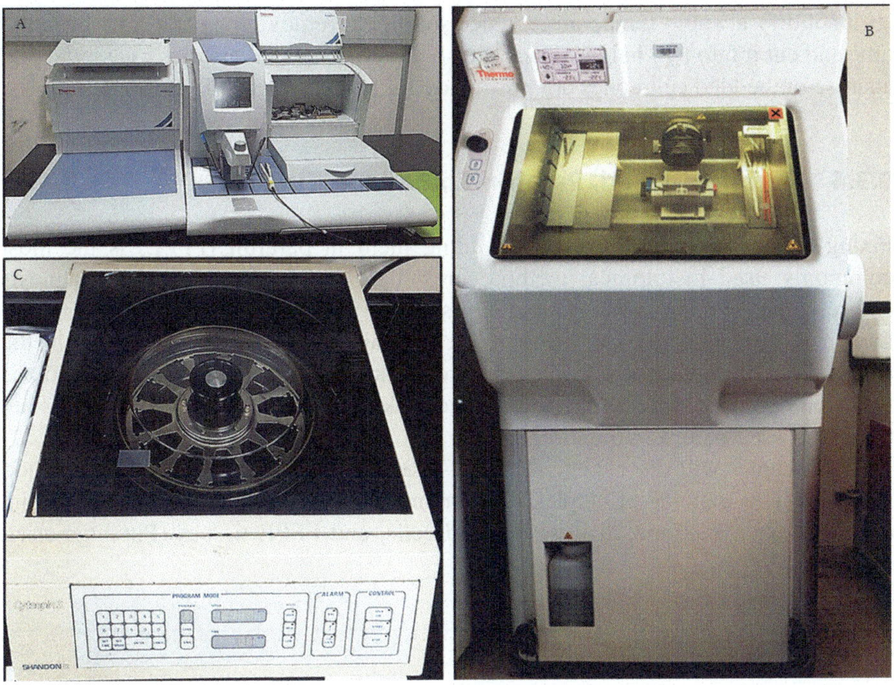

Fig. 9.1 Instruments for histological samples preparation. (**a**) FFPE embedding station. (**b**) Cryostat. (**c**) Cytospin

tissue architecture by stopping all molecular activities. Cryostat sample preparation is often used for quick diagnosis and for detecting clinically relevant antigens that are difficult to or cannot be detected in fixed tissues (Pittaluga et al. 2017). Frozen blocks for cryostat sectioning are prepared by embedding tissues in mounting media followed by quick immersion into liquid nitrogen. Cryostats are then used to cut the frozen blocks into sections thin enough for microscopic examination. It is recommended to fix frozen sections in cold acetone or alcohol-based fixatives before immunostaining.

9.3.3 Agarose-Blocked Cells

Laboratory-maintained cells can be processed for embedding in agarose to create solid physical support around them. This acts as an artificial extracellular matrix. To embed cells in agarose, wash cells and fix them in 10% formalin (4% paraformaldehyde) for at least 2 h at room temperature. After fixation, spin down the cells and discard the supernatants. Wash the cells in saline buffer (e.g., PBS). Resuspend the pellet in the residual saline buffer (~100 µL) by flicking the bottom of the tube. Gently add while vortexing ~1 mL of 2% molten agarose (at 50 °C), quick spin, and let it solidify at room temperature. Once solidified, remove the cells cone from the tube and cut it into two halves. Place each half into a labelled histology cassette. The agarose-embedded cell cones are now ready for processing as FFPE tissue samples.

9.3.4 Formalin-Fixed, Paraffin-Embedded Tissues

Fixing tissues in formalin and embedding them in paraffin (FFPE) is the most commonly used histological method for sample preparation, both for diagnosis and research. Here, a tissue sample is first preserved by fixing it in formalin (also known as formaldehyde) to preserve the proteins and vital structures within the tissue. Formalin preserves tissues by crosslinking proteins. Formalin fixation is followed by embedding in paraffin wax to make a histology block. The blocks are then processed before immunoreaction.

This chapter will discuss the formalin-fixed, paraffin-embedded (FFPE) method in-depth. The general principles of immunostaining are the same for cytospin, cryostat frozen sections, agarose-blocked cells, and FFPE tissues (Pittaluga et al. 2017).

9.4 Procedure

9.4.1 Processing of FFPE Blocks

FFPE histology blocks prepared from cells or tissue are processed as follows. **Note**: The term 'tissue' is subsequently used to refer to both agarose-blocked cells and tissue samples.

1. **Fixation and tissue processing.**
   (a) Fix the tissue (in the cassette) by incubating it in 10% formalin (4% paraformaldehyde) for 30 min at room temperature.
   (b) Dehydrate it in graded concentrations of ethanol (50%, 70%, 90%, 100% (x1), and 100% (x2)) for 30 min each at room temperature. (**Note**: Dehydration is critical as the presence of water in the tissue sample can interfere with the quality of the sample. Most tissues can tolerate overnight incubation in 70% ethanol at 4 °C without detrimental effects. Therefore, after initial dehydration in 70% ethanol for 30 min at room temperature, the tissues can be incubated in 70% ethanol at 4 °C overnight).
   (c) Gradually replace ethanol in the sample with xylene by incubating cassettes in an ethanol-xylene mixture (1:1), followed by incubating in 2 fresh absolute xylenes each for 30 min. (**Note**: Ethanol is not miscible with wax; therefore, it is essential to completely remove ethanol for the embedding wax to penetrate and occupy the spaces within the tissue sample.)
   (d) Impregnate the tissue in wax by incubating the cassettes in 3 fresh molten wax at 60 °C for 30 min each. (**Note**: After first incubation in wax for 30 min,

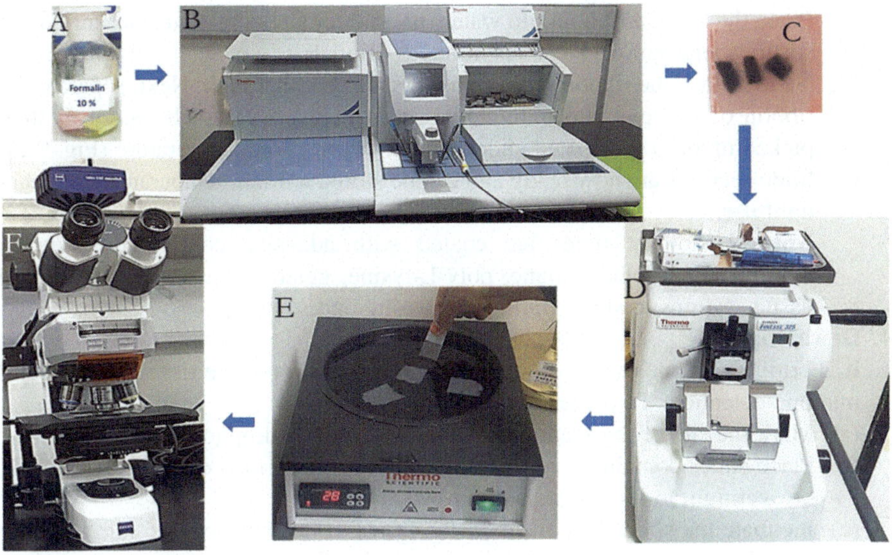

Fig. 9.2 Schematic representation of FFPE sample preparation and processing. (**a**) Tissues in 10% formalin. (**b**) FFPE embedding station. (**c**) FFPE tissue blocks. (**d**) Microtome for sectioning. (**e**) Picking sections in the water bath. (**f**) Microscope for visualising stained tissue

the cassettes can withstand storage at room temperature (overnight for cells, or up to 3 months for tissue samples) or proceed to the other two wax incubations for 30 min each. Now samples are ready for embedding.)

2. **Embedding.**
   (a) Place an embedding mould on the hot surface (60 °C) of an embedding facility, fill it with molten wax and then place the processed tissue into the molten wax (Fig. 9.2). (**Note:** It is important to use the same wax used in the impregnation for embedding to avoid problems during the dewaxing step.)
   (b) Transfer the mould to the cooled surface of the embedding facility. Quickly adjust the tissue position and orientation, then cover it with the cassette lid. Let it solidify.
   (c) Once solidified, cool the mould in a freezer (−20 °C) for 10 min (optional).
   (d) Remove the paraffin block from the mould. Scrape off excess wax on the sides of the block and ensure the cassettes are labelled.
   (e) Paraffin-embedded blocks are now ready for sectioning. FFPE blocks can be stored at room temperature for years.
3. **Sectioning.**
   Prior to sectioning, it is recommended to cool down the embedded blocks in a refrigerator for a few hours or overnight. This cooling helps prevent tissue damage and aids in cutting the sections in the microtone.
   (a) Place the block onto a microtome holder and trim off the wax by cutting ~10 to 15 μm until the tissue sample is reached (Fig. 9.2).
   (b) Cut samples into ~3 to 5 μm thick sections.

(c) Place the section ribbon into water (normal temperature water) to unfold the sections and remove wrinkles, if any (optional).
(d) Gently place the ribbon of sections in a 42 °C water bath. Next, separate the ribbon of sections into individual sections (Fig. 9.2). Each section is then picked up onto pre-coated slides and air-dry at room temperature (Fig. 9.2).
(e) Slide sections are now ready and can be stored at room temperature for years until use.
Note: Histology slides are coated with adhesive compounds such as 3-aminpropyltriethoxysilane, poly-L-lysine, gelatin, etc., to help retain the tissue sections to slides during processing and staining.

4. **Dewaxing.**
It is critical to remove wax from FFPE tissues to expose the tissue for probing/immune detection. To achieve this,
(a) Mark the site of the section on the slide with a hydrophobic barrier pen.
(b) Place slides containing sections in a 42 °C pre-warmed oven for 10–30 min or overnight to loosen the paraffin wax.
(c) Incubate the slides in 2 fresh changings of 100% xylene for 5 min each for complete dewaxing.
(d) Rinse the slides in graded ethanol (70%, 90%, and 100%) to remove residual xylene in the sections.
Note: If the final detection antibody is conjugated to horseradish peroxidase, endogenous peroxidase activity must be blocked.
(e) To block peroxidase activity, incubate the sections in 0.5% H_2O_2 (in methanol) for 20 min at room temperature, then rinse in 100% ethanol. (**Note**: This is only required when using the horseradish peroxidase detection method.)
(f) Incubate sections in a 37 °C pre-warmed water bath for 2 min.
(g) Add 0.1–10 mg/mL of proteinase K solution (in TE or PBS Tween) to a final volume of 100–200 μL per section to disrupt protein cross-links and expose the epitopes of the antigen.
(h) Incubate sections in proteinase K at 37 °C for 10–15 min.
(i) Rinse sections with tap water and then ascending graded ethanol (70%, 90%, and 100%).
(j) Air-dry the sections.

The sections are now ready for detection using ISH or immunoreaction for ICC. For IHC, however, additional processing may be required before immunoreactions. These include rehydration, antigen retrieval, and/or permeabilisation.

1. ***Rehydration*** is needed because most IHC staining solutions are aqueous. Rehydration is performed by incubating the sections in 2 fresh changings of 100% xylene for 5 min each, then in graded ethanol (100%, 90%, 70%, and 50%) for 10 min each. Finally, rinse the sections in water.
2. ***Antigen retrieval*** is done to unmask epitopes to facilitate the binding of antibodies. Although antigen retrieval is commonly used in IHC, it is only

essential for epitopes that are not otherwise detectable. It is performed using an enzymatic or heating method. The heating method involves boiling the slides for 10 min in 10 mM buffer, such as sodium citrate, sodium acetate, sodium phosphatase, or Tris-HCl, etc., with pH varying from 1 to 10 (4–8 is optimal for most of these buffers) (Shi et al. 1995). After heating, cool the slides at room temperature for about 30 min and then rinse in running water for 5 min. Enzymatic antigen retrieval by protease-induced method involves treating the sections with enzymes such as proteinase K, trypsin, or pepsin for 10–20 min at 37 °C. Then, rinse in running water for 5 min.

3. ***Permeabilisation*** is needed to facilitate the penetration of antibodies, especially if the target antigen is intracellular. Permeabilisation is achieved by incubating sections in permeabilisation buffer (1% animal serum and 0.4% Triton X-100 in PBS (or PBS Tween)) for 10 min and then rinsing in water for 5 min. The sections are now ready for antigen detection.

9.4.2 Antigen Detection

Viruses in histology sections are detected based on the antigen of interest and choice of technique. For example, in IHC and ICC, antigens are detected by immunoreaction, while in ISH, probing the nucleic acid of interest with a complementary probe is used for detection. Immunoreactions may follow probing in ISH. Multiple antigens can also be detected in a given section based on the compatibility of the antibodies used and the visualisation microscope. Below are procedures for some detection systems commonly used in histology laboratories.

1. **Nucleic acid probing using in situ hybridisation.**
   (a) Add appropriate concentration (0.1–1 μg/mL) of a labelled probe (specific to the viral nucleic acid of interest) to sections on slides.
   (b) Cover the sections with a coverslip and heat at a low-power microwave. (**Note**: This is important to unfold nucleic acid secondary structure and denature interacting proteins for efficient probe hybridisation.)
   (c) For hybridisation to occur, incubate the slides in a humid chamber at an appropriate temperature and time. For instance, 42 °C overnight was optimised for the detection of Epstein–Barr virus (EBV) using EBV-encoded RNA (EBER) probes (Khan et al. 1992) (Chap. 4, Fig. 4.2).
   (d) Wash the slides in saline buffers such as 2 × SSC for 5 min at room temperature to loosen the coverslip.
   (e) Flick the slides gently to remove the coverslips. (**Note**: Mechanical removal of the coverslip by hand may damage the sections).
   (f) Wash sections twice in 2 × SSC at room temperature for 5 min each.
   (g) Perform a stringency wash by incubating slides in 0.1 × SSC in a 50 °C water bath, shaking for 10–30 min. This wash removes any nonspecifically bound probe.

(h) Wash sections twice in 2 × SSC at room temperature and then 1 × PBS Tween wash for 5 min.
Now ISH sections are ready for immunoreactions or signal detection in case of directly conjugated probes.

2. **Immunoreaction.**
   (a) To the tissue section, add antibodies (can be monoclonal or polyclonal) specific to the viral antigen of interest (i.e., specific to the viral protein of interest in IHC and ICC or specific to the molecule used in labelling the probe in ISH).
   (b) Incubate the slide at an appropriate time and temperature for the antigen–antibody reaction. For instance, 1–2 h incubation at room temperature or overnight at 4 °C. Incubating sections overnight in diluted primary antibodies often provides better staining and less background compared to short time (1–2 h) incubation in highly concentrated antibodies. (**Note**: Antibodies can be directly conjugated with detection molecule (one-step reaction) or indirect (two-step reaction), requiring a secondary antibody specific to the Fc portion of the primary antibody. For more details, see Chap. 5, under types of ELISA.)
   (c) In the case of indirect methods, incubate the sections in appropriate secondary antibodies for 30–60 min at room temperature (**Note**: Prolong incubation may result in high background.)
   (d) Wash slides in 1 × PBS Tween twice for 5 min each.
   Now sections are ready for signal detection.
3. **Double staining.**
   (a) Perform immunoreaction with an antibody of interest (first antibody).
   (b) Develop the first antibody using the appropriate signal detection method, e.g., ABC-peroxidase/DAB (see the following section: Signal detection and visualisation).
   (c) Wash the sections in saline buffers.
   (d) Repeat antigen retrieval and blocking for the second time (optional).
   (e) Immunoblot the section with another antibody of interest (second antibody). The second antibody must target a different viral antigen than the first antibody. (**Note**: The second antibody must be conjugated with a different detection molecule than the one used with the first antibody. For example, if ABC-peroxidase/DAB detection system is used in the first round, then a different detection system, like alkaline phosphatase/NBT/BCIP can be used for the second antigen, Fig. 9.3).
   (f) Wash the sections in saline buffers.
   (g) Block in an appropriate blocking reagent.

Now slides are ready for signal detection and visualisation.

NOTE: If the detection substrate is fluorescent, the antibodies must be tagged to different fluorochromes. Here, the two antibodies can be incubated at the same time. After immunoreaction, wash the section and quench it to limit the background fluorescence (Fig. 9.3).

9.4.3 Signal Detection and Visualisation

Several detection methods are used in histology for the detection of viral antigens from tissue samples (Fig. 9.3). Each detection method uses a specific reagent. These reagents can be chromogenic, such as 3′-diaminobenzidine (DAB) and p-nitroblue tetrazolium chloride/5-bromo-4-chloro-3-indolyl phosphate (NBT/BCIP) or

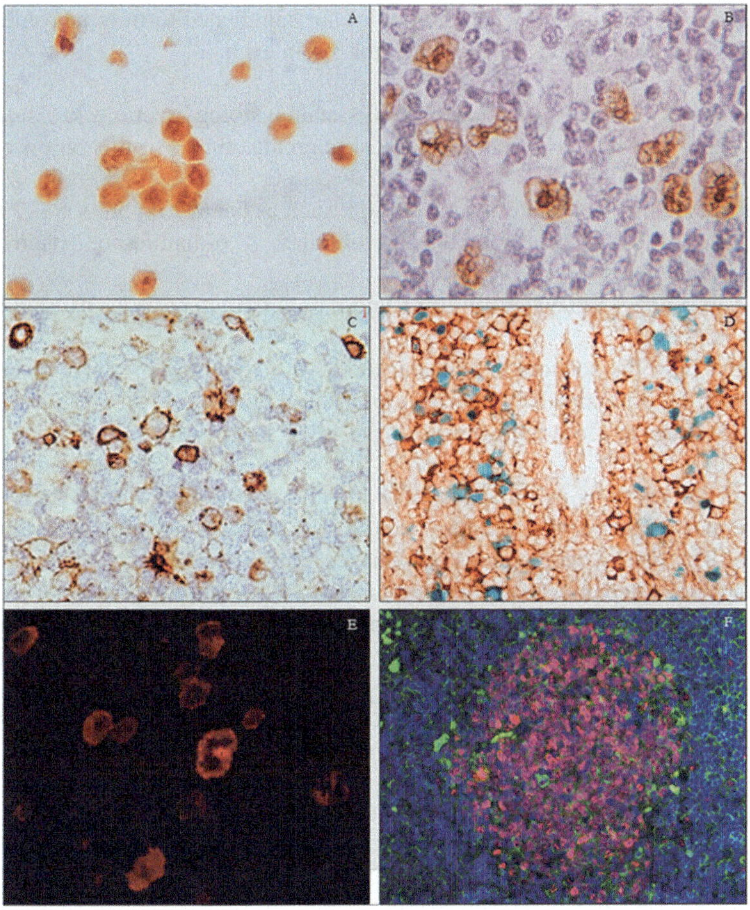

Fig. 9.3 Images of FFPE-stained sections detecting antigens of Epstein–Barr virus. (**a**) Detection of EBV using ISH in a cell line infected with EBV with a chromogenic single detection system. (**b**) Detection of EBV in a lymph node tissue from a patient with EBV-associated Hodgkin lymphoma with a chromogenic single detection system. (**c**) Detection of EBV latent membrane protein (LMP1) in a spleen tissue using IHC with a chromogenic single staining detection system. (**d**) Detection of EBV proteins, LMP1 and EBNA1 in a spleen tissue using IHC with chromogenic double staining detection system. (**e**) Detection of EBV LMP1 in a spleen tissue using IHC with a fluorescent single staining detection system. (**f**) Detection of proliferating B-cells in a spleen tissue using IHC with fluorescent double staining detection system

fluorescent, such as fluorescein isothiocyanate (FITC). This section will discuss the general procedure for signal detection in histology.

- Add ~200 μL of the detection reagent (e.g., ABC, alkaline phosphatase) per section and incubate it in a humidified chamber for 30 min at room temperature.
- Wash slides twice in 1 × PBS Tween for 5 min each.
- Develop the signal by incubating sections in appropriate substrates. For example, chromogens such as DAB for ABC, NBT/BCIP for alkaline phosphatase. (**Note**: This step is not required when antibodies are conjugated to fluorochromes.)
- Wash slides in 1 × PBS Tween twice for 5 min each.

Note: It is always good practice to counter-stain histology sections to visualise the negative cells and appreciate the overall tissue architecture. For this purpose.

- Incubate slides in a counter-stain solution, such as haematoxylin, for 1–2 min. In the case of fluorescently labelled antibodies, 4′,6-diamidino-2-phenylindole (DAPI) can be used.
- Rinse in running tap water for 2 min and then differentiate in 1–3% acid alcohol for 1 min.
- Finally, blue the sections by rinsing them with running tap water for 10 min, followed by 90% and 100% ethanol for 1 min each.
- Air-dry slides at room temperature.

Mount the sections in an appropriate mountant to protect them from physical damage and preserve the staining for more extended storage. Mounting also improves the clarity and contrast of images during microscopy. The choice of mounting media depends on the detection method used. Some examples of mounting media are DPX xylene-based, and flouromountants aqueous-based. To mount.

- Add mounting media to the section.
- Cover the sections with a coverslip such as a 50 mm glass coverslip.
- View under a microscope (light or fluorescent).

Troubleshooting tips are highlighted (Table 9.1).

9.5 Troubleshooting

Table 9.1 Troubleshooting for histology

S/N	Problem	Possible reason(s)	Solution(s)
1.	No signal	• Inappropriate primary antibody • Incompatibility of primary or secondary antibodies • Probe-antibody incompatibility (in ISH) • The antigen is masked	• Check the specificity of antibodies • Ensure compatibility of primary and secondary antibodies • Ensure probe-antibody compatibility and concentration • Perform and optimise antigen retrieval conditions
2.	Nonspecific signals	• Cross-reactivity of antibodies • Nonspecific antibody binding	• Check the specificity of probes/antibodies • Optimise probe/antibody concentration and incubation time • Increase time and/or salt concentration or rocking speed during stringency wash
3.	High background noise	• Nonspecific probe/antibody binding • Nonspecific antibody binding	• Optimise probe/antibody concentration and incubation time • Increase temperature in stringency washes • Increase salt concentration in wash buffers • Quench immunofluorescence slides to reduce autofluorescence • Optimise probe/antibody concentration and incubation time • Increase time and rocking speed during washes

References

Brunzel NA (2016) Body fluid analysis. In: Fundamentals of urine and body fluid analysis. Elsevier Health Sciences, pp 356–358

Hassani A, Reguraman N, Shehab S, Khan G (2021) Primary peripheral Epstein-Barr virus infection can lead to CNS infection and neuroinflammation in a rabbit model: implications for multiple sclerosis pathogenesis. Front Immunol 12:1

Hopkins PM et al (2003) Indirect fluorescent antibody testing of nasopharyngeal swabs for influenza diagnosis in lung transplant recipients. J Heart Lung Transplant 22:161–168

Khan G, Coates PJ, Kangro HO, Slavin G (1992) Epstein Barr virus (EBV) encoded small RNAs : targets for detection by in situ hybridisation with oligonucleotide probes. J Clin Pathol 45:616–620

Pittaluga S, Barry TS, Raffeld M (2017) Immunohistochemistry for the hematopathology laboratory. In Hema 41–52.e4

Reguraman N, Hassani A, Philip P, Khan G (2021) Uncovering early events in primary Epstein-Barr virus infection using a rabbit model. Sci Rep 11:21220

Renshaw S (2017) Immunohistochemistry and immunocytochemistry. In: Immunohistochemistry and immunocytochemistry. John Wiley & Sons, Ltd, pp 35–102. https://doi.org/10.1002/9781118717769.ch3

Shi SR, Imam SA, Young L, Cote RJ, Taylor CR (1995) Antigen retrieval immunohistochemistry under the influence of pH using monoclonal antibodies. J Histochem Cytochem 43:193–201

Treuhaft MW, Soukup JM, Sullivan BJ (1985) Practical recommendations for the detection of pediatric respiratory syncytial virus infections. J Clin Microbiol 22:270–273

Bioinformatics and In Silico Stimulations 10

The development of genomics and high-throughput experimental technologies such as sequencing creates the need for the use of computers to store and analyse the large amounts of data generated. Therefore, bioinformatics was developed as a discipline. It is associated with sequence databases and sequence analysis—thus, a computational science using biological data. Although there are several definitions and views on bioinformatics, most researchers now use the term 'bioinformatics' as a generic term for both the storage and maintenance of biological data and the use of computational data analysis and algorithms in studies related to functional genomics. Bioinformatics involves several scientific fields including computer science, mathematics, statistics, informatics, physics, chemistry, biology, and medicine.

Biologists worked closely with their computer science counterparts to put together meaningful biological information into computer algorithms for easy and convenient analyses and communications. Bioinformatics is essential for analysing large molecular data, system biology, biotechnology, legal and forensic techniques. Before the advent of bioinformatics, biological experiments were only performed either in vivo (inside cell) or in vitro (in glass). With the advent of bioinformatic field, biological systems are now studied in silico (on silicon chip microprocessors) using computer algorithms. Therefore, bioinformatics can be defined as the computational branch of molecular biology.

In molecular biology, bioinformatics is important in genomic data analysis that involves several vital steps such as genomic sequence assembly, genome mapping, gene prediction, gene annotation and comparative genomics, and identification of new strains or characteristics of microbes, including viruses. Genomic DNA, RNA, plasmids, and proteins are now being studied using computer software and applications such as BLAST, BioEdit, Biology workbench, etc., for in silico designs and de novo assembly. This chapter focused on sequence analysis tools.

Z. Hassan, G. Khan, *Molecular Techniques for Studying Viruses*,
https://doi.org/10.1007/978-981-99-8097-0_10

10.1 Sequencing and Sequence Analysis

Sequencing is the process of determining the nucleotide (or amino acid) order of a given macromolecule. This is achieved by labelling the nucleotides using fluorescent dyes and visualising them (Chap. 9, Fig. 9.3). Several macromolecules can be sequenced. These include.

- **DNA:** DNA encodes for the necessary information required for a living organism to survive and reproduce. Determining the DNA sequence of an organism reveals useful information fundamental to the organism's life and evolution. There are many techniques for DNA sequencing. One of the common approaches is sequencing a fragment(s) of the gene or genome. Whole genome (or long DNA segments) is also sequenced. This is achieved by first dividing the genome into smaller fragments, then sequencing each of the fragments separately, and finally assembling the resulting genomic sequence. There are various browsers available for genomic sequence analyses. These include the University of California, Santa Cruz (UCSC) Human Genome Browser, NCBI, Ensembl, etc. These browsers have the same genome sequences; however, they could differ in their presentation designs and graphics for annotation.
- **RNA:** RNA encodes for the protein-coding region of the genome. It is transcribed from DNA by an enzyme—RNA polymerase. **Note**: Some viruses have RNA as their genome. These viruses either transcribe their mRNA from their RNA genome or undergo reverse transcription. The sequence of an RNA can be determined through traditional sequencing techniques and high-throughput RNA sequencing (RNA-Seq). RNA-Seq is a specialised high-throughput sequencing method that provides insight into the whole cellular transcriptome. Most of the browsers used in DNA sequence analysis can be used for analysing RNA sequences.
- **Protein:** Protein is the product obtained from translating the coding region of RNA. Protein sequence is determined by sequencing the amino acid of its polypeptide or the whole protein. Proteins are sequenced directly using methods such as Edman degradation and spectroscopy. Protein sequence can also be obtained via in silico translation of nucleic acids sequences. Primary protein sequences or in silico translated sequences can be analysed using BLAST algorithms, BioEdit, ExPASy, SIB Bioinformatics portal (https://www.expasy.org), etc. ExPASy is a user-friendly interface that provides several protein analyses including MSA, system and structural analyses.

10.2 Genetic Relatedness and Phylogenetic Analysis

Phylogenetic analysis is a process of testing the evolutional relatedness of living organisms using their biological characteristics (Persing et al. 2011). These characteristics include physical traits such as shape, temperature requirements, pathologies, and virulence. In recent years, molecular characteristics including

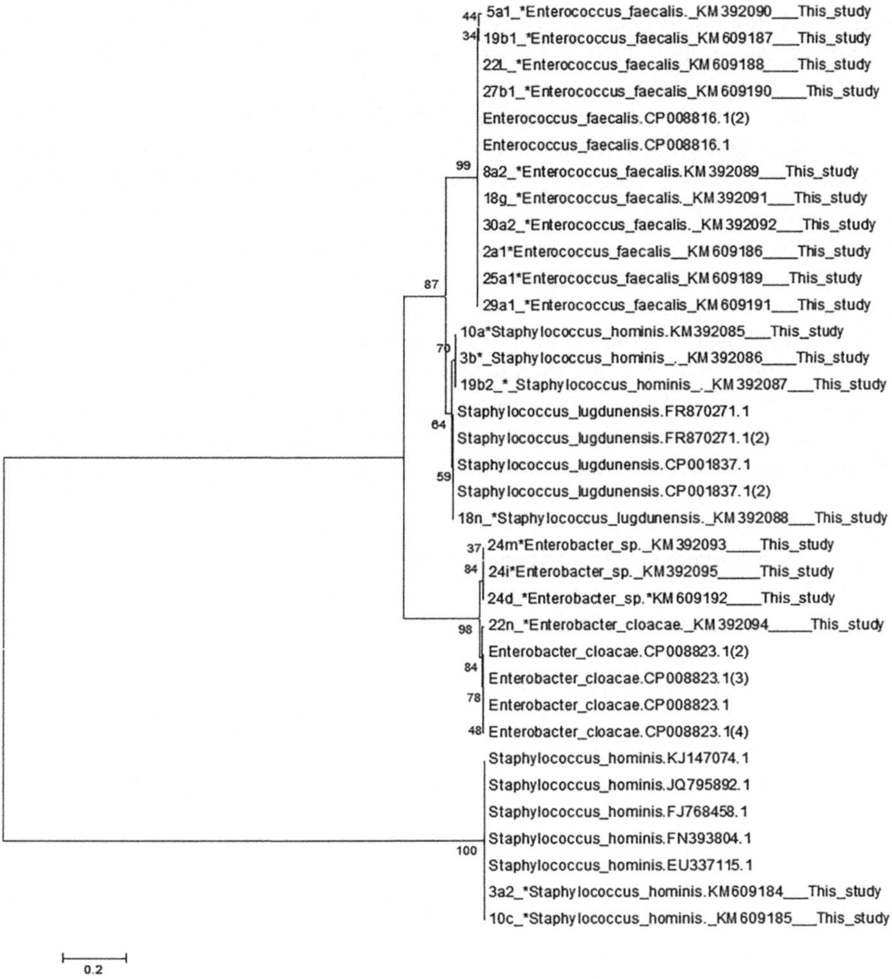

Fig. 10.1 Phylogenetic tree of 16S rDNA gene of bacteria isolated from human milk

DNA, RNA, genes, and protein are frequently being used in depicting phylogenetic relationships (Fig. 10.1). Phylogenetics is important in understanding how molecular sequences evolve. Furthermore, it enables the prediction of how species evolve and how they will change in the future (Bush 2001). In the diagnostic field, phylogenetic analysis is used in classifying and characterising pathogens, including viruses. Moreover, it is essential in forensic studies and identification of new pathogens.

Sequences used to construct a phylogenetic tree are often edited and curated (Hassan et al. 2016). The sequences can be obtained from primary data or retrieved from online sources, such as DNA Data Bank of Japan (DDBJ), Japan (http://www.ddbj.nig.ac.jp), European Nucleotide Archive (ENA), EMBL-EBI, UK (http://www.ebi.ac.uk), International Nucleotide Sequence Database (INSD) (http://www.insdc.

org/), and the GenBank, NCBI, USA (http://www.ncbi.nlm.nih.gov/genbank). Moreover, data from both primary and online sources can be combined in a single tree for comparison of their genetic relatedness. Recent advancements in bioinformatics make the use of phylogeny prediction tools available. The tools include several algorithms such as neighbour joining, maximum likelihood, maximum parsimony, etc (Saitou and Nei 1987). and software such as BioEdit™ and molecular evolutionary genetics analysis (MEGA) (Tamura et al. 2007). Statistical parameters used to test for the robustness of the tree or its individual branches include sequence divergence, and bootstrapping with a certain number of replications (e.g. 1000) (Felsenstein 1985) (Fig. 10.1). Some of the phylogeny tools are equipped with several visual aids for better visualisation and annotation.

Phylogenetic tree showing the relative positions of isolates from human breast milk as inferred by the neighbour-joining method of partial 16S rDNA sequences. Percentage of bootstrap values for a total of 1000 replicates are shown at the nodes of the tree. Sequence divergence was at 0.2 scales (Hassan et al. 2016).

10.3 Macromolecules Interactions

10.3.1 Structural Predictions

Once the sequence of a molecule is known, there are many easy-to-follow web servers available for structural predictions. These algorithms are governed by different statistical measures including minimum energies of formation, F-value, and propensity scores (in the range of 0 to 1 **or** − 1 to +1). **Note:** For a beginner, the use of the default parameters of these tools is advised.

Here are some websites available for structural prediction:

1. *DNA structure:* RDNAnalyzer, mfold, DNAproDB web server, computational recognition of secondary structure (CROSS).
2. *RNA structure:* RNAfold web server, mfold, and CROSS.
3. *Protein structure:* NovaFold web server, AlphaFold website, RoseTTAFold website, etc.

10.3.2 In Silico Macromolecular Interactions

Like structural predictions, there are online tools available for predicting macromolecular interactions. These tools are also governed by different statistical measures including, distinct nonredundant benchmark datasets classifiers, support vector machine (SVM), random forest (RF) (Muppirala et al. 2011; Sanchez de Groot et al. 2019), area under the curve (AUC), receiver operating characteristic (ROC) (Muppirala et al. 2011), Fisher's exact test for combinatorial complexity of the reference RNA, Z-score, propensity scores dissociation constant (Kd), binding free energy (ΔG), etc. Most of these parameters are ranged between 0 and 1 or − 1

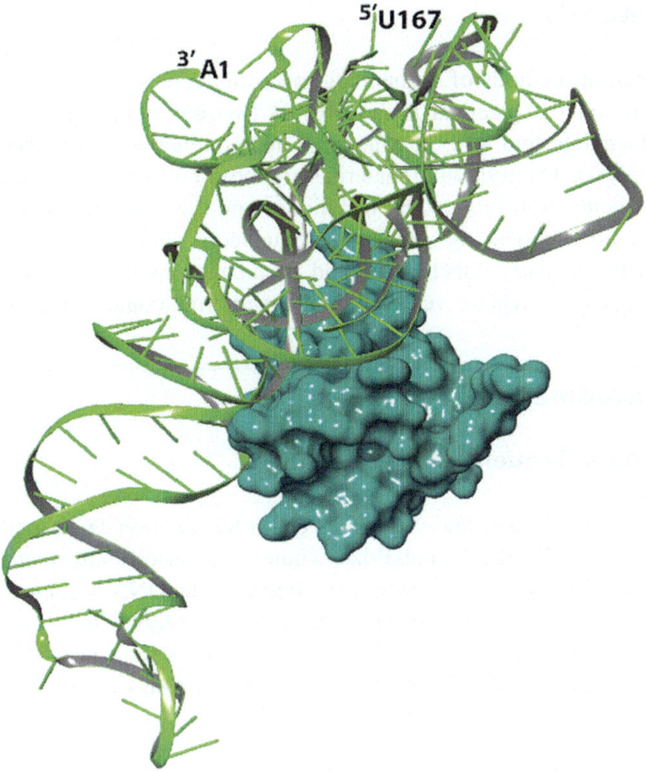

Fig. 10.2 In silico prediction of RNA-protein interaction

and + 1. An example of nucleic acid– protein prediction output is shown in Fig. 10.2. Below are some of the tools used for predicting macromolecular interaction.

1. *DNA–protein interaction:* ProDFace, ProNA2020, etc.
2. *RNA–protein interaction:* ProNA2020, RNA–protein interaction using sequence only (RPISeq) (http://pridb.gdcb.iastate.edu/RPISeq/), HDOCK (http://hdock.phys.hust.edu.cn/), catRAPID omics.
3. *Protein–protein interaction:* TRI_tool, ProNA2020, etc.

In silico prediction of the interaction Epstein-Barr virus-encoded small RNA-1 (green) with human Lupus La N-terminal domain (cyan) carried out using HDOCK (http://hdock.phys.hust.edu.cn/).

10.4 Materials

There are several methods of sequencing macromolecules. However, in this book, we will focus on the practical procedures of two common sequencing methods, Sanger sequencing and the next-generation sequencing techniques, and their sequence analysis. The following materials are required to perform Sanger sequencing and next-generation sequencing techniques and the sequence analysis.

Nucleic acid, sequencing reagents, sequencer, agarose gel electrophoresis reagents, PCR reagents, ddNTPs, 5′ and 3′ adapter molecules, magnetic stand, sonicator, sequencing reactor vessel, computer, internet connection, software.

10.5 Procedure

10.5.1 Sanger Sequencing Technique

Sanger sequencing is also known as the 'chain termination PCR method'. It was developed in 1977 by the Nobel Prize winner, Frederick Sanger and colleagues (Sanger et al. 1977). Sanger sequencing technique uses DNA sequence of less than 1000 bp nucleotides in length. It has approximately 99.99% base accuracy and is thus considered the 'gold standard' for validating DNA sequences (De Cario et al. 2020; Franklin et al. 2014). There are mainly two versions of Sanger sequencing—the classical method and the automated method.

10.5.1.1 Classical Sanger Sequencing Method

The DNA fragment to be sequenced is amplified using PCR protocol (as described in Chap. 4 of this book). The PCR product is then purified and used as template for the sequencing (Fig. 10.3).

Preparation of Sequencing Mixture

Prepare the sequencing mixture by adding the purified PCR products, primers complementary to the strand to be sequenced, DNA polymerase, the four dNTPs, and one dideoxynucleotide triphosphates (ddNTPs)—(ddATP, ddGTP, ddCTP, and ddTTP—one type of ddNTP per tube). **Note**: ddNTPs are added in a low ratio compared to the dNTPs. Now, subject the sequencing mixture to the following steps for sequencing.

Sequencing Thermocycling

- Heat the sequencing mixture at 94–98 °C for 1–2 min to denature the dsDNA in the PCR product into two ssDNA.
- Lower the heat to the optimum primer annealing temperature for 0.5–2 min. This will allow the primer to bind to its complementary region at the 5′ end of the DNA strand.
- The temperature is maintained at 30–50 °C for DNA polymerase to extend the strand until a termination nucleotide (ddNTP) is randomly incorporated.

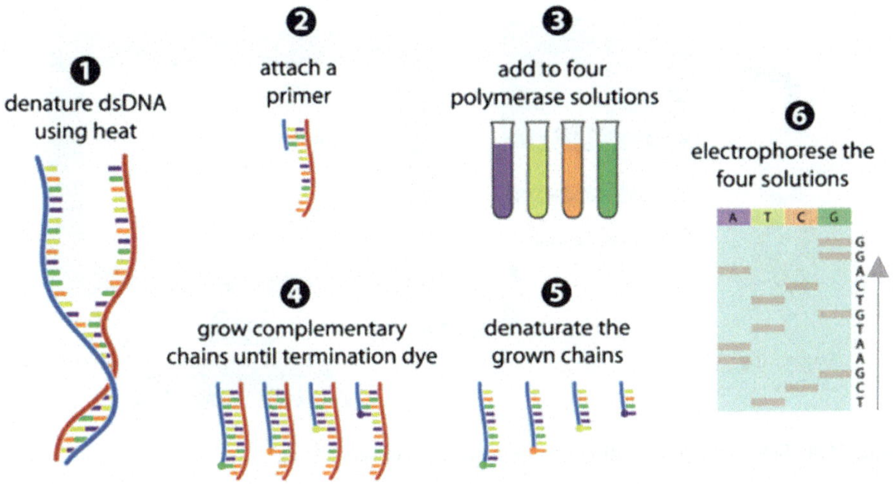

Fig. 10.3 Schematic presentation of the principles of classical Sanger sequencing

- The thermocycling is repeated for 30–40 cycles.
- The reaction is finally heated at 60–80 °C to release the extension fragments.

Analysis and Determination of DNA Sequence

- The resulting ssDNA fragments are separated by gel electrophoresis.
- All four ddNTP reaction tubes shall be separated on the same gel with a reaction per lane.
- The sequence is determined by the length of the band on the gel, starting the read from the bottom of the gel to the top (i.e., from smallest to largest band).

Note: In classical Sanger sequencing, four tubes should be prepared, each with only one type of ddNTP added (Fig. 10.3). Once a ddNTP is added, polymerase activity stops as it cannot catalyse the formation of a phosphodiester bond between ddNTP and the 5′-phosphate of the next nucleotide.

10.5.1.2 Automated Sanger Sequencing Method

Preparation of Sequencing Mixture

Prepare the sequencing mixture by adding the purified PCR products, primers complementary to the strand to be sequenced, DNA polymerase, the four dNTPs, and four ddNTPs—(ddATP, ddGTP, ddCTP, and ddTTP—all four types of ddNTPs are added in one sequencing tube) (Fig. 10.4). **Note**: Each ddNTP type should be labelled with a unique fluorescent dye and is added in low ratio compared to the dNTPs. Now, subject the sequencing mixture to the following steps for sequencing.

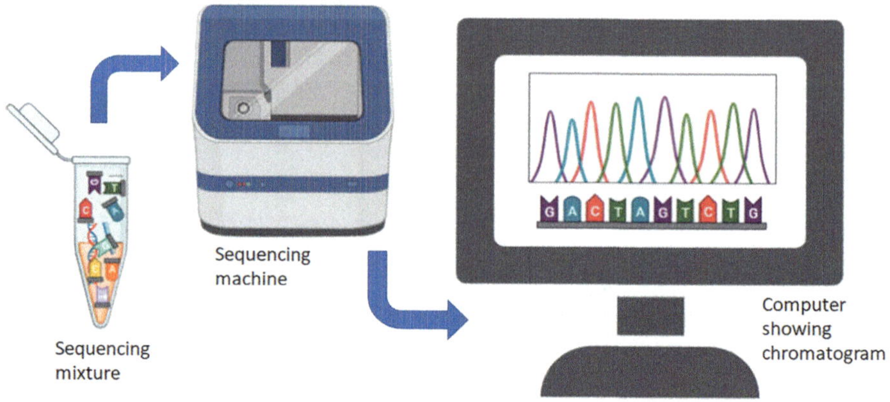

Fig. 10.4 Schematic presentation of automated Sanger sequencing

Sequencing Thermocycling

- Heat the sequencing mixture at 94–98 °C for 1–2 min to denature the dsDNA in the PCR product into two ssDNA.
- Lower the heat to the optimum primer annealing temperature for 0.5–2 min. This will allow the primer to bind to its complementary region at the 5′ end of the DNA strand.
- The temperature is maintained at 30–50 °C for DNA polymerase to extend the strand until a termination nucleotide (ddNTP) is randomly incorporated.
- The thermocycling is repeated for 30–40 cycles.
- The reaction is finally heated at 60–80 °C to release the extension fragments.

Analysis and Determination of DNA Sequence

- All oligonucleotides are run in a single capillary electrophoresis chamber in the sequencing machine, for 2–3 h.
- With the help of laser excitement, the fluorescent label attached to the ddNTPs emits light that is detected by a computer. The computer then reads each band of the capillary gel in order.
- The nucleotide sequence is visualised as a chromatogram, which shows the fluorescent peak of each nucleotide along the length of the template DNA.

10.5.2 Next-Generation Sequencing Technique

Next-generation sequencing (NGS) was developed as a modification of the traditional Sanger sequencing technique. NGS is based on sequencing-by-synthesis (SBS) technology and uses reversible dye terminators that enable the identification of single bases as they are introduced into DNA strands (Fig. 10.5). There are three basic steps in performing NGS. These include.

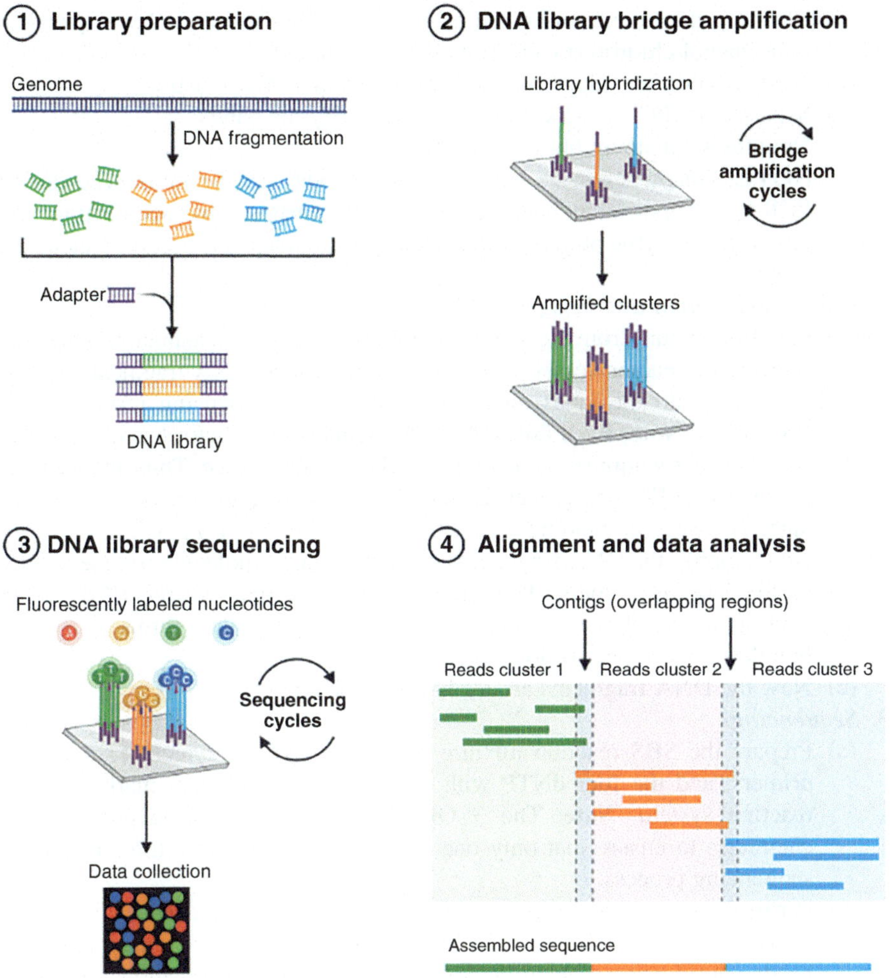

Fig. 10.5 Schematic presentation of next-generation sequencing

1. ***Library preparation.***
   (a) The DNA to be sequenced is fragmented to around 200–500 bp in length (Fig. 10.5) by ultrasonic fragmentation or enzymatic restriction, e.g., using transposase.
   (b) Prepare the adapter by phosphorylating the 5′ end adapter and tailing the 3′ end adapter.
   (c) Add adapter to the fragmented DNA in ~10:1 ratio adapter: fragment and incubate at 55–65 °C for 20 min. The time can be extended up to 1 h depending on the protocol. These will allow the adapters to bind to the ends of these small DNA segments.

   (d) Clean up the excess un-ligated adapters by either magnetic beads purification, phenol-chloroform, ethanol precipitation, etc.
   (e) Next, amplify the adapter-ligated fragments in a PCR reaction.
   (f) Separate the PCR products on a gel and purify the bands.
   (g) Now the sequencing library is ready.
   NOTE: Adapters can be barcoded, or barcodes can be introduced during PCR by using different barcoded primers. If transposase is used for fragmentation, the enzyme automatically adds specific adapters to both ends of the fragments.
2. ***Cluster generation and bridge amplification.***
   (a) Pass the mixture from the sequencing library through a channel into the core sequencing reactor vessel. Here the adapter-ligated DNA fragments will be adsorbed onto the lanes on the surface of the sequencing reactor vessel (Fig. 10.5). **Note**: Each flow cell on the sequencing reactor has 8 lanes, and each lane has a number of adapters attached to the surface. Thus, the adapters ligated to the DNA fragments bind to their matched numbers on the flow cell lane. This specific binding is thus termed as '*building process*'.
   (b) Now amplify the DNA fragments using the adapter molecules on the surface of the lane as primers. This amplification is called '*bridge PCR*'. After continuous amplification cycles, each DNA fragment forms cluster in bundles at its respective locations.
   (c) Now the DNA fragments are ready for sequencing.
3. ***Sequencing.***
   (a) Prepare the SBS reaction mixture by adding DNA polymerase, connector primers, and the four dNTP with base-specific fluorescent markers in the reaction system. **Note**: The 3′-OH of the dNTP must be protected by chemicals to ensure that only one base will be added at a time during the sequencing process.
   (b) Perform the sequencing for 4–24 h, based on the system used.
   (c) Remove all unused free dNTP and DNA polymerase by either magnetic beads purification, phenol-chloroform, ethanol precipitation, or other methods.
   (d) Add buffering solution for fluorescence excitation. The fluorescence is then excited by laser and its signal is recorded by the optical equipment attached to a computer.
   (e) The optical signal is converted into the sequencing base by the computer for analysis. **Note**: After the fluorescence signal is recorded, a quencher can be added to quench the fluorescence signal and remove the dNTP 3′-OH protective group, so that another round of sequencing reaction can be performed.
   (f) Now align the cluster sequences to obtain a consensus sequence of the DNA.

10.5.3 Sequence Analysis

Sequencing results are generated in chromatograms. Often, some positions tend to produce noise that could result in generating unclear nucleotide(s) at that position. Therefore, the sequencer generates possible nucleotide based on the in-build algorithms. Table 10.1 presents sequencing symbols and their interpretation. A single fragment is often sequenced more than once to increase reliability of the data. Therefore, the sequences are aligned in a multiple sequence alignment (MSA) tool to generate a contig sequence. Here, we will demonstrate MSA using ClustalW in BioEdit™ software (Hall 1999).

10.5.3.1 Steps of Blasting and Retrieving Sequences from NCBI Website

- Input, edit, and trim sequences for quality assurance using BioEdit™ software.
- Perform MSA of the multiple sets of forward and the reverse complement of the reverse sequences (in case of double-stranded molecule) to generate a consensus sequence using the Contig assembly programme.
- BLAST the consensus sequence online for highly similar sequences to identify the organism.

Note: 'blastn' algorithm is used for searching nucleotide database using nucleotide sequences, 'blastp' algorithm is for searching protein database using protein sequences, 'blastx' algorithm is for searching protein database using translated nucleotide sequences, 'tblastn' algorithm is for searching translated nucleotide sequences database using protein sequences, 'tblastx' algorithm is for searching translated nucleotide database using translated nucleotide sequences.

Table 10.1 Interpreting sequencing symbols

S/N	Nucleotide symbol	Nucleotide name
1.	A	Adenine
2.	C	Cytosine
3.	G	Guanine
4.	T	Thymine
5.	U	Uracil
6.	R	Guanine/adenine (purine)
7.	Y	Cytosine/thymine (pyrimidine)
8.	K	Guanine/thymine
9.	M	Adenine/cytosine
10.	S	Guanine/cytosine
11.	W	Adenine/thymine
12.	B	Guanine/thymine/cytosine
13.	D	Guanine/adenine/thymine
14.	H	Adenine/cytosine/thymine
15.	V	Guanine/cytosine/adenine
16.	N	Any nucleotide

NOTE: Sequence search can be restricted to a specific genus or reference sequence (Hassan et al. 2016; Zhao et al. 2015).

10.5.3.2 Steps in Using the NCBI Website to Search and Retrieve Sequences

- Input your query sequence in the NCBI website.
- Scroll down and click the 'BLAST' button to search.
- Check for the similarity percentage of the query sequence and those from the NCBI website using various formats, such as the E-value, percentage identity, graphical view, alignment view etc.
- Next, select the most similar sequences and download them in the 'FASTA aligned' format. This will download only the segment of the sequences that align with the query sequence (Fig. 10.6).
- Now the sequences can be used in MSA and phylogenetic analyses (Fig. 10.7).

10.5.3.3 Steps of Sequence and Phylogenetic Analyses in BioEdit™ Software

- Launch the BioEdit™ software and input the query sequences (sequence data from sequencing or those retrieved from NCBI).
- Select all sequences to be included in the MSA by dragging the sequence name (from the left panel).
- Select and run ClustalW multiple sequence alignment from the 'accessory application' pull-down menu.
- View the MSA by selecting any preferred viewing format.

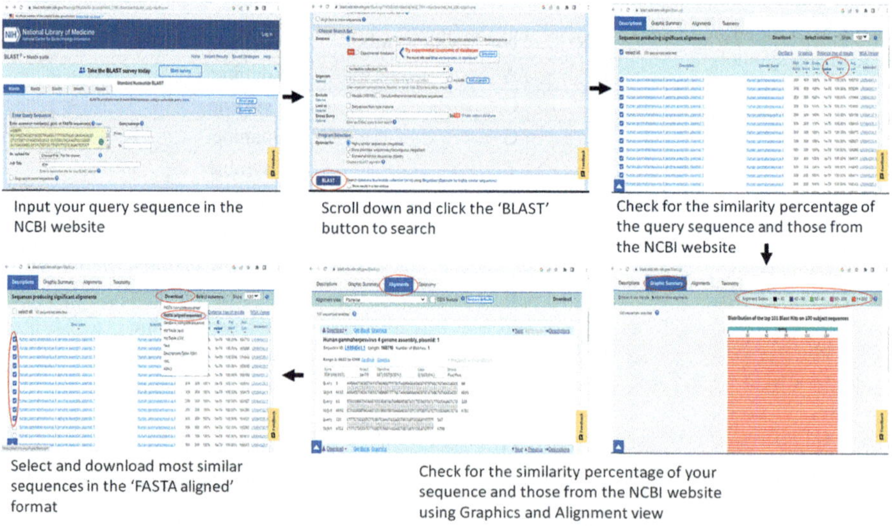

Fig. 10.6 BLAST and retrieving sequences from NCBI website

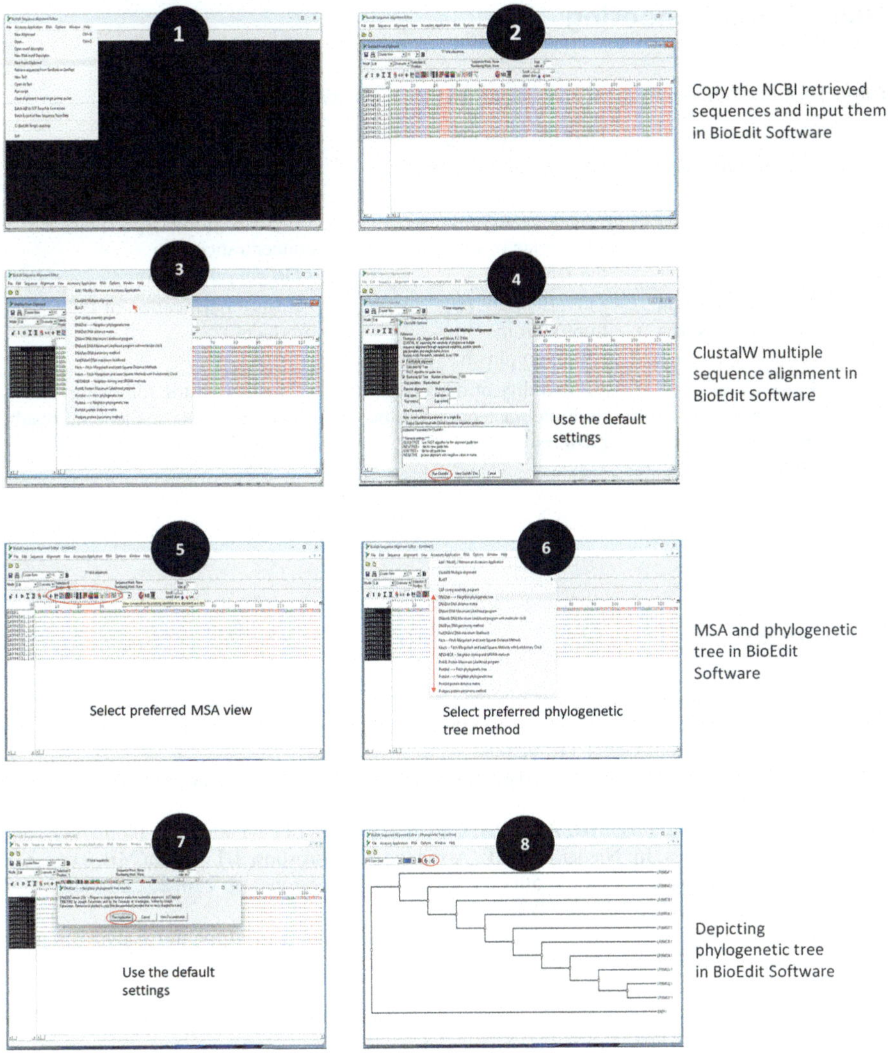

Fig. 10.7 Sequence and phylogenetic analyses in BioEdit™ software

- To depict the phylogenetic relationship of the sequence data, select all sequences to be included in the phylogenetic analysis by dragging the sequence name (from the left panel).
- Select and run preferred phylogenetic algorithm, e.g., neighbour joining, from the 'accessory application' pull-down menu.
- View the phylogenetic tree by selecting any preferred viewing format (Fig. 10.7).

Troubleshooting tips are highlighted (Table 10.2).

10.6 Troubleshooting

Table 10.2 Troubleshooting for sequencing and sequence analysis

S/N	Problem	Possible reason(s)	Solution(s)
1.	No or poor signal	• No or poor quality of DNA	• Increase DNA concentration • Perform DNA amplification
		• Inadequate primers	• Increase primer specificity and concentration
		• Presence of impurities in the DNA	• Do gel purification
2.	Multiple peaks at the same position	• Overlapping of fluorochrome • Polymorphism	• Ensure purification of PCR products
3.	Presence of unusual nucleotides alphabet	• Possibility of detecting multiple nucleotides at the position	• Presence of the following alphabets in the sequencing result is normal

References

Bush RM (2001) Predicting adaptive evolution. Nat Rev Genet 2:387–392

De Cario R et al (2020) Sanger validation of high-throughput sequencing in genetic diagnosis: still the best practice? Front Genet 11:592588

Felsenstein J (1985) Confidence limits on phylogenesis: an approach using the bootstrap. Evolution 39:1596–1599

Franklin WA, Aisner DL, Post MD, Bunn PA, Garcia MV (2014) 17—Pathology, biomarkers, and molecular diagnostics. In: Niederhuber JE, Armitage JO, Doroshow JH, Kastan MB, Tepper JE (eds) Abeloff's clinical oncology, 5th edn. Churchill Livingstone, pp 226–252.e6. https://doi.org/10.1016/B978-1-4557-2865-7.00017-5

Hall TA (1999) BioEdit: a user-friendly biological sequence alignment editor and analysis program for windows 95/98/NT. Nucleic Acids Symp Ser 41:95–98

Hassan Z, Mustafa S, Abdul Rahim R, Isa NM (2016) Identification, characterisation and phylogenetic analysis of commensal bacteria isolated from human breast milk in Malaysia. Pertanika J Sci Technol 24:351–370

Muppirala UK, Honavar VG, Dobbs D (2011) Predicting RNA-protein interactions using only sequence information. BMC Bioinformatics 12:489

Persing DH et al (2011) Molecular microbiology diagnostic principles and practice. ASM Press, Washington, DC

Saitou N, Nei M (1987) The neighbor-joining method: a new method for reconstructing phylogenetic trees. Mol Biol 4:406–425

Sanchez de Groot N et al (2019) RNA structure drives interaction with proteins. Nat Commun 10:1–13

Sanger F, Nicklen S, Coulson AR (1977) DNA sequencing with chain-terminating inhibitors. Proc Natl Acad Sci U S A 74:5463–5467

Tamura K, Dudley J, Nei M, Kumar S (2007) MEGA4: Molecular Evolutionary Genetics Analysis (MEGA) software version 4.0. Mol Biol Evol 24:1596–1599

Zhao J et al (2015) High-throughput sequencing of RNAs isolated by cross-linking immunoprecipitation (HITS-CLIP) reveals argonaute-associated microRNAs and targets in Schistosoma japonicum. Parasit Vectors 8:589

Summary and Conclusion

11

11.1 Summary

Infectious diseases are among the leading cause of premature deaths (Graham and Sullivan 2018). More than 30 new human pathogens have been identified in the last three decades, most of which are of zoonotic origin (Greger 2007). The increasing number of zoonotic diseases is alarming and probably reflects the close human–animal interface (Graham and Sullivan 2018; Greger 2007; Hassani and Khan 2020) and the increase in international travel (Institute of Medicine (US) 2010). Viruses, in particular, are the most important zoonotic pathogens and a constant public health threat (Morens et al. 2004). The current COVID-19 pandemic and the devastating impact such outbreaks can have are evident (Jones 2020). Therefore, studying viruses is essential for routine viral research and rapid detection and diagnosis of these infections. Molecular approaches are arguably the best viable options in this regard. With molecular techniques, viruses and viral components can be detected within a short time and from minimalist samples.

Furthermore, there have been enormous advances in molecular biology, genetics, and biotechnology in the last few decades. These advances have led to unprecedented technical developments in genomics, transcriptomics, proteomics, and imagining technologies. It is now possible to detect, interrogate, and map viral infections by analysing thousands of markers simultaneously in the context of tissue morphology. These tissue microarrays and spatial transcriptomics provide insights into the biology and pathogenesis of viral diseases not previously possible. These technologies, however, are expensive and not accessible for routine practical use. Therefore, low-cost and effective methods, like those described in this book, are needed to get students familiar with such techniques.

Z. Hassan, G. Khan, *Molecular Techniques for Studying Viruses*,
https://doi.org/10.1007/978-981-99-8097-0_11

11.2 Conclusion

Molecular laboratory techniques are essential methods for viral studies and promising tools for innovations in virology. Therefore, understanding the fundamental principles of these techniques and how they work is important for correct data interpretation. Moreover, understanding the principles allows for modification of a protocol to tailor it for one's laboratory setup. Furthermore, technical advancement in these methods based on situational context could increase a proactive approach to targeting viral pathogens locally and thus reduce over-dependency on diagnosis, vaccinology, and treatments.

References

Graham BS, Sullivan NJ (2018) Emerging viral diseases from a vaccinology perspective: preparing for the next pandemic. Nat Immunol 19:20–28

Greger M (2007) The human/animal interface: emergence and resurgence of zoonotic infectious diseases. Crit Rev Microbiol 33:243–299

Hassani A, Khan G (2020) Human-animal interaction and the emergence of SARS-CoV-2. JMIR Public Health Surveill 6:e22117

Institute of Medicine (US) (2010) Forum on microbial travel, conflict, trade, and disease. Infectious disease movement in a borderless World: workshop summary. National Academies Press, Washington, DC

Jones DS (2020) History in a crisis—lessons for covid-19. N Engl J Med 382:1681–1683

Morens DM, Folkers GK, Fauci AS (2004) The challenge of emerging and re-emerging infectious diseases. Nature 430:242–249

Index

Z. Hassan, G. Khan, *Molecular Techniques for Studying Viruses*,
https://doi.org/10.1007/978-981-99-8097-0

GPSR Compliance

The European Union's (EU) General Product Safety Regulation (GPSR) is a set of rules that requires consumer products to be safe and our obligations to ensure this.

If you have any concerns about our products, you can contact us on ProductSafety@springernature.com

In case Publisher is established outside the EU, the EU authorized representative is:

Springer Nature Customer Service Center GmbH
Europaplatz 3
69115 Heidelberg, Germany

Batch number: 10738117

Printed by Printforce, the Netherlands